APPLICATIOɴ OF
DEEP LEARNING METHODS IN
HEALTHCARE AND MEDICAL SCIENCE

APPLICATION OF DEEP LEARNING METHODS IN HEALTHCARE AND MEDICAL SCIENCE

Edited by
Rohit Tanwar, PhD
Prashant Kumar, PhD
Malay Kumar, PhD
Neha Nandal, PhD

First edition published 2023

Apple Academic Press Inc.
1265 Goldenrod Circle, NE,
Palm Bay, FL 32905 USA

760 Laurentian Drive, Unit 19,
Burlington, ON L7N 0A4, CANADA

CRC Press
6000 Broken Sound Parkway NW,
Suite 300, Boca Raton, FL 33487-2742 USA

4 Park Square, Milton Park,
Abingdon, Oxon, OX14 4RN UK

Library and Archives Canada Cataloguing in Publication

Title: Application of deep learning methods in healthcare and medical science / edited by Rohit Tanwar, PhD, Prashant Kumar, PhD, Malay Kumar, PhD, Neha Nandal, PhD.

Names: Tanwar, Rohit, editor. | Kumar, Prashant, 1986- editor. | Kumar, Malay, 1988- editor. | Nandal, Neha, 1990- editor.

Description: First edition. | Includes bibliographical references and index.

Identifiers: Canadiana (print) 20220285055 | Canadiana (ebook) 20220285071 | ISBN 9781774911204 (hardcover) | ISBN 9781774911211 (softcover) | ISBN 9781003303855 (ebook)

Subjects: LCSH: Medical technology. | LCSH: Medical care—Technological innovations. | LCSH: Deep learning (Machine learning)

Classification: LCC R855.3 .A66 2023 | DDC 610.285—dc23

Library of Congress Cataloging-in-Publication Data

Names: Tanwar, Rohit, editor. | Kumar, Prashant, 1986- editor. | Kumar, Malay, 1988- editor. | Nandal, Neha, 1990- editor.

Title: Application of deep learning methods in healthcare and medical science / edited by Rohit Tanwar, Prashant Kumar, Malay Kumar, Neha Nandal.

Description: First edition. | Palm Bay, FL : Apple Academic Press, 2023. | Includes bibliographical references and index. | Summary: "This volume provides a wealth of up-to-date information on developments and applications of deep learning in healthcare and medicine. It aims to provide deep insight and understanding of novel applications that address the tough questions of disease diagnosis, prevention, and immunization. The volume looks at applications of deep learning for major medical challenges such as cancer detection and identification, birth asphyxia among neonates, kidney abnormalities, white blood cell segmentation, diabetic retinopathy detection, and Covid-19 diagnosis, prevention, and immunization. The volume discusses applications of deep learning in detection, diagnosis, intensive examination and evaluation, genomic sequencing, convolutional neural networks for image recognition and processing, and more for health issues such as kidney problems, brain tumors, lung damage, and breast cancer. The authors look at ML for brain tumor segmentation, in lung CT scans, in digital X-Ray devices, and for a logistic and transport systems for effective delivery of healthcare. Chapters include studies and discussions on chest X-ray images using CNN to identify Covid-19 infections, lung CT scan images using pre-trained VGG-16 and 3-layer CNN to distinguish Covid and non-Covid patients, genomic sequencing to study the Covid virus, breast cancer identification using CNN, brain tumor detection using multimodal image fusion and segmentation, factors responsible for birth asphyxia in neonates, and much more. It also explores cancer identification and detection using deep learning methods in the human body through algorithms based on issues, laboratory tests, imaging tests, biopsies, bone scans, computerized tomography scans, positron emission tomography, and ultrasound. This volume, Application of Deep Learning Methods in Healthcare and Medical Science, showcases the diverse applications of patient-based data collection and analysis in medicine and healthcare using computer algorithms for effective health diagnosis, prevention, and patient care"-- Provided by publisher.

Identifiers: LCCN 2022031830 (print) | LCCN 2022031831 (ebook) | ISBN 9781774911204 (hardcover) | ISBN 9781774911211 (paperback) | ISBN 9781003303855 (ebook)

Subjects: MESH: Medical Informatics | Deep Learning

Classification: LCC R855.3 (print) | LCC R855.3 (ebook) | NLM W 26.55.A7 | DDC 610.285--dc23/eng/20220824

LC record available at https://lccn.loc.gov/2022031830

LC ebook record available at https://lccn.loc.gov/2022031831

ISBN: 978-1-77491-120-4 (hbk)
ISBN: 978-1-77491-121-1 (pbk)
ISBN: 978-1-00330-385-5 (ebk)

About the Editors

Rohit Tanwar, PhD
Associate Professor, School of Computer Sciences,
University of Petroleum and Energy Studies, Dehradun,
Uttarakhand, India, Tel.: +91-9992257914,
E-mail: rohit.tanwar.cse@gmail.com

Rohit Tanwar, PhD, is an Associate Professor at the School of Computer Sciences, University of Petroleum and Energy Studies, Dehradun, India. He has more than 10 years of experience in teaching. His areas of interests include network security, optimization techniques, human computing, soft computing, cloud computing, data mining, etc. Dr. Tanwar has published several books in the domain of healthcare and security with reputed international publishers. He is associated with some highly indexed international journals as guest editor/onboard reviewer. He has more than 30 publications to his credit to date in reputed journals and conferences. He has been associated with many conferences throughout India as member, session chair, etc. He is supervising PhD research scholars in the fields of security and optimization. Dr. Tanwar received his bachelor's degree in CSE from Kurukshetra University, Kurukshetra, India, and master's degree in CSE from YMCA University of Science and Technology, Faridabad, India. He has received his PhD in CSE from Kurukshetra University, India.

Prashant Kumar, PhD
Assistant Professor, Department of Computer Science
and Engineering, Dr. BR Ambedkar National Institute of
Technology, Jalandhar, Punjab, India,
Tel.: +91-8351976199, E-mail: prashantk@nitj.ac.in

Prashant Kumar, PhD, is an Assistant Professor in the Department of Computer Science and Engineering at the Dr. B. R. Ambedkar National Institute of Technology, Jalandhar, India. Previously he has worked with the Department of Systemics in the School of Computer Science at the

University of Petroleum and Energy Studies, Dehradun, India, and the Department of Computer Science and Engineering at the National Institute of Technology Hamirpur, India. His research interests include opportunistic and delay-tolerant networks, device-to-device communications, wireless and adhoc networks, and security in wireless networks. He has published more than 25 research papers in journals and conferences of international repute. He is a member of IEEE, International Association of Engineers, and the Internet Society. Dr. Kumar received his PhD and MTech degrees from the National Institute of Technology Hamirpur, India.

Malay Kumar, PhD
Assistant Professor, Department of Computer Science and Engineering, Indian Institute of Information Technology, Dharwad IT Park, Hubli, Karnataka, India, Tel.: +91-89595-96477,
E-mail: malay.kumar@iiitdwd.ac.in

Malay Kumar, PhD, is an Assistant Professor in the Department of Computer Science and Engineering at the Indian Institute of Information Technology Dharwad, India. Earlier he was associated with the School of Computer Science of the University of Petroleum and Energy Studies, Dehradun, India. He has authored more than 20 research papers in international journals and conferences. His research areas of interest are application of machine learning and deep learning in medical sciences, and security and privacy issues in cloud computing. He has served as chair and technical program committee member for numerous international conferences and workshops. He was a guest editor of several international journals and a lead editor of several books. Dr. Kumar earned his BTech in Computer Science and Engineering from CSJM University Kanpur, his MTech from NIT Kurukshetra, and his PhD from NIT Raipur, India.

Neha Nandal, PhD
Associate Professor, Department of Computer Science and Engineering, Gokaraju Rangaraju Institute of Engineering and Technology, Hyderabad, Telangana, India, Tel.: +91-8368959558, E-mail: neha28nandal@gmail.com

Neha Nandal, PhD, is an Associate Professor in the Computer Science and Engineering Department of the Gokaraju Rangaraju Institute of Engineering and Technology, Hyderabad, India. She has published 19 articles in her research area in different journals and conferences, including SCI- and SCOPUS-level journals. She is a life-time member of IETA. Dr Neha has participated in different workshops; completed courses on Python, machine learning, and deep learning on Coursera; and also hosted different faculty development programs. Her research interests include pattern recognition and machine learning. She earned her BTech in CSE from the Technological Institute of Textile and Sciences, Bhiwani, and her MTech in CSE from Amity University Jaipur, India, with distinction. Recently, she has been awarded a PhD in Machine Learning.

Contents

Contributors

Sakshi Aggarwal
Department of Computer Science and Engineering, Motilal Nehru National Institute of Technology Allahabad, Prayagraj, Uttar Pradesh, India, E-mail: sakshiaggarwal@mnnit.ac.in

Neelu Jyothi Ahuja
Department of Systemics, School of Computer Sciences, University of Petroleum and Energy Studies, Dehradun, Uttarakhand, India, E-mail neelu@ddn.upes.ac.in

Janibul Bashir
Department of Information Technology, National Institute of Technology, Srinagar, Jammu and Kashmir, India, E-mail: janibbashir@nitsri.ac.in

Saswati Kumari Behera
Department of Electrical and Electronics Engineering, Sri Sairam Engineering College, Chennai, Tamil Nadu, India, E-mail: saswatikumari.eee@sairam.edu.in

Chandradeep Bhatt
CSED, Graphic Era Hill University, Dehradun, Uttarakhand, India, E-mail: bhattchandradeep@gmail.com

Arvind Dhaka
Department of Computer and Communication Engineering, Manipal University Jaipur, Rajasthan, India, E-mail: arvind.neomatrix@gmail.com

R. Dhanalakshmi
IIIT Tiruchirappalli, Trichy, Tamil Nadu, India, E-mail: r_dhanalakshmi@yahoo.com

Todor Ganchev
Technical University of Varna, Varna, Bulgaria

Jitendra Kumar Gupta
CSED, GRD Institute of Management and Technology, Dehradun, Uttarakhand, India, E-mail: jk760429@gmail.com

Tahir Javed
Department of Information Technology, National Institute of Technology, Srinagar, Jammu and Kashmir, India

G. Kannan
Department of Electronics and Communication Engineering, B. S. Abdur Rahman Crescent Institute of Science and Technology, Vandalur, Chennai, Tamil Nadu, India

Rama Rao Karri
Brunei University of Technology, Brunei Darussalam, E-mail: karri.rao@utb.edu.bn

Sandeep Chand Kumain
CSED, Graphic Era Hill University, Dehradun, Uttarakhand, India, E-mail: skumain@gehu.ac.in

Ghost Manoj Kumar
Department of Electronics and Communication Engineering, Kallam Haranadha Reddy Institute of Technology, Chowdavaram, Guntur, Andhra Pradesh, India,
E-mail: Manojkumar.ece2016@gmail.com

Indrajeet Kumar
CSED, Graphic Era Hill University, Dehradun, Uttarakhand, India, E-mail: erindrajeet@gmail.com

P. Sathish Kumar
Department of Electronics and Communication Engineering, Bharath Institute of Higher Education and Research, Selaiyur, Chennai-73, Tamil Nadu, India, E-mail: sathishmrl30@gmail.com

K. K. Mishra
Department of Computer Science and Engineering, Motilal Nehru National Institute of Technology Allahabad, Prayagraj, Uttar Pradesh, India

Amita Nandal
Department of Computer and Communication Engineering, Manipal University Jaipur, Rajasthan, India

Sheema Parwaz
Department of Computer Science Engineering, SVIET, Chandigarh, India

G. Prashanth
Software Engineer, Wipro Technology, Bangalore, Karnataka, India

V. Rajendran
Department of Electronics and Communication Engineering, Vels Institute of Science, Technology, and Advanced Studies (VISTAS), Chennai, Tamil Nadu, India

M. Ramamoorthy
Department of Artificial Intelligence and Machine Learning, Saveetha School of Engineering, SIMATS, Chennai, Tamil Nadu, India

Sweeti Sah
Department of Computer Science and Engineering, National Institute of Technology Puducherry, India, E-mail: sweetisah3@gmail.com

Gopal Sakarkar
G. H. Raisoni College of Engineering, Nagpur, Maharashtra, India,
E-mail: gopal.sakarkar@raisoni.net

Arpit Kumar Sharma
Department of Computer and Communication Engineering, Manipal University Jaipur, Rajasthan, India

Navjot Singh
Department of Information Technology, Indian Institute of Information Technology Allahabad, Prayagraj, Uttar Pradesh, India

Vemu Santhi Sri
Department of Electronics and Communication Engineering, Kallam Haranadha Reddy Institute of Technology, Chowdavaram, Guntur, Andhra Pradesh, India, E-mail: vemu.santhisri@gmail.com

B. Surendiran
Department of Computer Science and Engineering, National Institute of Technology Puducherry, Karaikal, India, E-mail: surendiran@nitpy.ac.in

Vivek Tiwari
IIIT Naya Raipur, Chhattisgarh, India, E-mail: vivek@iiitnr.edu.in

M. Padma Usha
Department of Electronics and Communication Engineering, B. S. Abdur Rahman Crescent Institute of Science and Technology, Vandalur, Chennai, Tamil Nadu, India, E-mail: padmausha@crescent.education

Amit Verma
School of Computer Science, UPES, Dehradun, Uttarakhand, India

G. Vijaya
Department of Computer Science and Engineering, Bapatla Women's Engineering College, Bapatla, Andhra Pradesh, India, E-mail: viji.pooshan@gmail.com

P. Vijayalakshmi
Department of Electronics and Communication Engineering, Vels Institute of Science, Technology, and Advanced Studies (VISTAS), Chennai, Tamil Nadu, India, E-mail: viji.se@velsuniv.ac.in

Siti Sophiayati Yuhaniz
Universiti Teknologi Malaysia, Johor Bahru, Malaysia, E-mail: sophia@utm.my

Abbreviations

3D	three-dimensional
ACO	ant colony optimization
AD	Alzheimer's disease
AE	autoencoder
AI	artificial intelligence
ALTS	autonomous logistic transportation system
AML	acute myeloid leukemia
ANN	artificial neural networks
AR	augmented reality
BL	Bayesian learning
CAC	coronary artery calcium
CAD	computer-aided diagnosis
CADe	computer-aided detection
CADT	computer-aided diagnostic tools
CAKUT	congenital abnormalities kidney and urinary tract
CAT	computer-aided tomography
CBA	cloud-based applications
CBC	complete blood cell
CGAN	conditional generative adversarial networks
CKD	chronic kidney disease
CL	convolution layer
CMSA	consensus multiple sequence alignments
CNN	conventional neural network
COVID-19	coronavirus disease
CSF	cerebrospinal fluid
CT	computed tomography
CVDs	cardiovascular diseases
CXR	chest X-ray
DAEs	denoising autoencoders
DICOM	digital imaging and communications in medicine
DL	deep learning
DM	data mining
DM	diabetes mellitus

DNN	deep convolutional neural networks
DR	diabetic retinopathy
DS	decision surface
DTs	decision trees
DWT	discrete wavelet transform
eGFR	estimated glomerular filtration rate
EHRs	electronic health records
EMR	electronic medical record
FMG	first moment gradient
FP	false positive
GAN	generative adversarial network
GC	guanine-cytosine
GDA	gradient descent algorithm
GLM	general linear model
GPUs	graphics processing units
HIPAA	Health Insurance Portability and Accountability Act
HM	hemorrhages
HMM	hidden Markov model
HMRF	hidden Markov random field
IOU	intersection over union
KFUH	King Fahd University Hospital
KNN	K-nearest neighbor
LR	linear regression
MA	microaneurysms
MCT	mobile computed tomography
MDCT	multi-detector computed tomography
MI	medical imaging
MLP	multilayer perceptron
MO	morphological opening
MRI	magnetic resonance imaging
MSA	multiple sequence alignment
NB	naïve Bayes
NN	neural network
NPDR	non-proliferative
NSP	nonstructural proteins
ORF	open reading frame
PA	poster anterior
PACS	picture archiving and communications system

PET	positron emission tomography
PPE	personal protective equipment
PPV	positive protected value
PSO	particle swarm optimization
RAT	rapid antigen test
RBCs	red-blood-cells
RF	random forest
RMSprop	root mean square propagation
RNA	ribonucleic acid
RNNs	recurrent neural networks
ROI	region of interest
RT-PCR	reverse transcription-polymerase chain reaction
SAE	stacked-autoencoder
SGD	stochastic gradient descent
SMG	second moment gradient
SPECT	single-photon emission computed tomography
SSIM	structural similarity index measure
SVM	support vector machine
US	ultrasound
VR	virtual reality
WBCs	white-blood-cells
WDBC	Wisconsin diagnostic breast cancer
WFH	work-from-home
WHO	World Health Organization
WM	white matter

Preface

The main objective of this book is to provide deep insight and understanding of novel applications of deep learning (DL) methods in healthcare and medical science. This book is a collection of work that addresses the tough questions of the global COVID pandemic in terms of disease diagnosis, prevention, and immunization. Additionally, the book covers other major medical challenges such as different types of cancer detection and identification, birth asphyxia among neonates, kidney abnormalities, white blood cell segmentation, and diabetic retinopathy (DR) detection. Another important addition in the book is a work on logistics and transport systems for effective delivery of healthcare.

The chapters of this book cover a comprehensive study of chest X-ray (CXR) images using CNN to identify COVID-19 infections and a study on lung CT scan images using pre-trained VGG-16 and 3-layer CNN to distinguish COVID and COVID patients. Further, the book discusses DL-driven diagnostic solutions for COVID prevention, diagnostics, and immunization.

Further, the book includes device development for a non-invasive COVID-19 identification procedure and a DL technique that is integrated with a portable X-ray device. The X-ray devices captured the X-ray images of the chest of suspected COVID patients and identified the infected patient. Finally, the book includes a comprehensive discussion on genomic sequencing to study the COVID virus. This study helps to discover medicine and is vital to effectively treat patients.

Further, the book identifies solving other medical problems using DL methods. It addresses breast cancer identification using CNN, and the book also dedicates a chapter to cancer identification and detection using DL methods in the human body through an algorithm based on issues, laboratory tests, imaging tests, biopsy, bone scan, computerized tomography scan, positron emission tomography (PET), and ultrasound.

Furthermore, the book also includes a work on brain tumor detection using multimodal image fusion and segmentation. The book also presents a study on critical factors responsible for birth asphyxia in neonates. This study helps pediatricians by including the therapeutic measures to use. An important contribution in the book is a detailed review of state-of-the-art methods that focus on digital imagery to understand kidney abnormalities. Additionally, analysis of white blood cells helps medical practitioners to understand the underlying disease conditions of patients. This book presents a detailed study of white blood cells using DL methods.

Finally, the book covers an important study on the detection of diabetic retinopathy (DR). A DL-based framework is designed to automate the detection of DR and provide identification of its seriousness.

CHAPTER 1

A Review on Detection of Kidney Disease Using Machine Learning and Deep Learning Techniques

VEMU SANTHI SRI,[1] P. SATHISH KUMAR,[2] and V. RAJENDRAN[3]

[1]Department of Electronics and Communication Engineering, Kallam Haranadha Reddy Institute of Technology, Chowdavaram, Guntur, Andhra Pradesh, India, E-mail: vemu.santhisri@gmail.com

[2]Department of Electronics and Communication Engineering, Bharath Institute of Higher Education and Research, Selaiyur, Chennai-73, Tamil Nadu, India, E-mail: sathishmrl30@gmail.com

[3]Department of Electronics and Communication Engineering, Vels Institute of Science, Technology, and Advanced Studies (VISTAS), Chennai, Tamil Nadu, India

ABSTRACT

The incredible accomplishment of deep learning models at image identification responsibilities in modern existence traverse through an occasion of significantly improved utilization of electronic medicinal reports and problem-solving imaging. Currently, kidney disease is an emergent issue that spreads pandemic. As a result of the high possibility of death within a very short duration, kidney patients necessities being liverish and suitably made therapy. In this review, we surveyed several kidney disease identifications like tumors, calculi, cysts, and kidney stones using machine learning algorithms such as support vector machine, Naïve Bayes (NB),

Application of Deep Learning Methods in Healthcare and Medical Science.
Rohit Tanwar, PhD, Prashant Kumar, PhD, Malay Kumar, PhD, & Neha Nandal, PhD (Editors)
© 2023 Apple Academic Press, Inc. Co-published with CRC Press (Taylor & Francis)

classification, and neural network (NN) were utilized to investigate the accuracy of performance in kidney disease prediction. Also, comparison of kidney disease prediction using deep learning techniques were analyzed to apply in remedial image analysis, spotlight on Convolutional NN, as well as highlight medicinal features of the field. This chapter gives a detailed literature review on the diagnosis of kidney abnormalities from various types of images and winds up by arguing research hindrances, emerging fields of machine learning, and deep learning approaches.

1.1 INTRODUCTION

Chronic kidney disease (CKD) is otherwise known as chronic kidney failure, which depicts the ongoing failure of kidney functioning. In other words, waste from the kidney, as well as surplus fluids from blood, that is emitted in your urine. CKD arises once either sickness or situations damage the kidney usage leads to kidney damage to deteriorate more than numerous months or years. This CKD arises because of the following conditions:

• The patients have type-I and type-II diabetes;
• Also, high blood pressure;
• Polycystic kidney disease.

This review focused on how kidney disease was predicted and also distinguished the persons who is affected by disease via images like mammogram, CT scan, and US images from normal using artificial intelligence techniques by several researchers.

The main objective of this review work is:

• to study various machine learning algorithms in detecting kidney diseases such as cysts, tumors, and calculi.
• to analyze how deep learning (DL) algorithms are applicable in finding abnormalities in the kidney, which may be either cysts or calculi, or tumors.
• to compare machine learning and DL-based NN algorithms to recognize which approaches are suitable in CKD diagnosis.
• to contrast the accuracy measures of various algorithms to evaluate the overall performance of algorithm in disease prediction.

- to examine several architectures of CNN algorithm developed by several researchers to highlight the disease prediction which helpful in medicinal applications.

1.1.1 DIAGNOSING KIDNEY DISEASE (SPECIFIC TUMOR)

Justin et al. [1] introduced how DL algorithms are applicable in medical diagnoses, such as image analysis, classification, localization, detection, image segmentation, and registration as well. DL-based convolutional neural network (CNN) is mainly valuable in medical image analysis since CNN maintains spatial interaction while input images are passing through filters like the convolutional layer, ReLU layer, pooling layer, fully connected layer, and finally output layer.

This DL-based CNN suitable in diagnosing several diseases in medicinal applications. Here, for instance, kidney disease classification task has been done through CNN to classify the input image as kidney tumor and kidney non-tumor depicted in Figure 1.1.

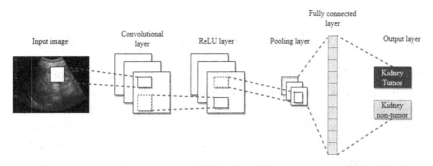

FIGURE 1.1 In this instance kidney disease classification task, an input image of an abnormal classification done through CNN.

Nour et al. [2] proposed a new DL model-based BPSO-DT (Binary Particle Swarm Optimization with Decision Tree (DT) along with CNN to categorize various tumors twisted in several parts of the body, especially kidney tumors. Classification of kidney tumor categories were predicted using layers in CNN depends on input ultrasound (US) images.

1.2   TECHNIQUES TO DETECT KIDNEY DISEASE

1.2.1   *DEEP LEARNING APPROACH*

Cao et al. [3] introduced how recent technologies enlargement, such as DL and neural network (NN) are applicable in the medical domain comprising of image segmentation, analysis, classification, prediction, and optimization.

1.2.1.1   *FINDING KIDNEY ABNORMALITIES, ESPECIALLY LESION*

Hui et al. [4] established a tool which is to detect injury (lesion) in kidney through many intersections over union (IOU) threshold depends on morphological CNN. Also, this proposal anticipated two morphological Convolutional layers as well as feature pyramid networks in faster RCNN and integration of four IOU threshold cascade RCNN. In this proposal, the finding of lesions in kidney using CT images as input reaches greater accuracy as 84%.

1.2.1.2   *KIDNEY DISEASE PREDICTION DEPENDS ON ATTRIBUTES*

Hu et al. [5] In spite of feature extraction of malicious behavior recognition, various kinds of malicious finding, the basic flow of detecting abnormal behavior structure is depicted in Figure 1.2. First training video, preprocessing, which undergoes feature reduction, then extraction of features, modeling event, model detection, and behavior classification is to categorize malicious from normal behavior. The training accuracy is somewhat higher than testing accuracy through experiment outcomes. Based on accuracy estimation, the performance of finding abnormal behavior is evaluated.

Inês Domingues et al. [6] introduced a survey regarding utilization of DL frameworks in contesting images, presence of dissimilar image modalities: first thing dynamically calculated computed tomography (CT) second thing is Positron emission tomography (PET), and also amalgamation of both modalities, significant to find several diseases. In this review, the researcher analyzed nearly 180 relevant studies published among 2014 to 2019 for the purpose of imaging modality.

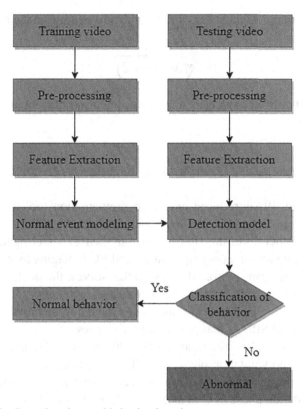

FIGURE 1.2   Detecting abnormal behavior detection.

Vinayagam et al. [7] this work is related to nephrolithiasis through Magnetic Resonance Images as input image which is preprocessed via Discrete Wavelet Transform (DWT). The feature extractions have been done by means of gray level co-occurrence matrix. This investigator surveyed 20 test data comprised of normal and abnormal MRI kidney images are categorized using NN, especially back-propagation. For image segmentation as kidney tumor or kidney stone, fuzzy clustering mean algorithm is used.

By Huang et al. [8], the number of hidden neurons available in two hidden layer is determined using the following equation:

$$\sqrt{(m+2)N} + \sqrt{2\left(\frac{N}{M+2}\right)} \tag{1}$$

Eqn. (1) being applicable to find a number of neurons in the first hidden layer.

$$\sqrt[m]{N / (m+2)} \tag{2}$$

Eqn. (2) utilized to identify a number of neurons present in the second hidden layer.

1.2.1.3   KIDNEY DISEASE DIAGNOSIS USING SEGMENTATION

Jackson et al. [9] introduced an image segmentation tool that depends on 3D CNN to distinguish left, right kidney shape by getting CT images as input. Furthermore, the shapes were applied to voxel quantity chart computed from post treatment quantitative SPECT imaging to approximate renal radiation quantity from therapy. In this survey, the model trained 89 contour cases as well as tested on same cases manually. Computerized contouring with CNNs provides assurance related to quantitative evaluation of practical SPECT and probably PET images.

Timothy et al. [10] investigated total 2,000 patients MR images utilized for training and validation phase, also 400 cases utilized for testing. This survey examined that DL-based multi-observer ensemble approach applied for segmentation phase as well as computation of total kidney volume for poly cysts kidney disease patients.

1.2.1.4   PREDICTING KIDNEY DISEASE BY USING CLINICAL TESTS

Vijaya et al. [11] introduced the DL model applied in Renal biopsy tests, which were gathered from 171 cases take care at Boston Medical Center from the period 2009 to 2012. DL-based CNN models were trained using images as inputs also training classes as outputs correspondingly.

1.2.1.5   PREDICTING CHRONIC KIDNEY DISEASE

Abdullah et al. [12] investigated DL models like Feed Forward NNs, also Wide and Deep models to make a diagnosis of CKD, performance calculation via metrics such as F1-score, precision, recall, accuracy prediction too. In addition to that, wide, and deep model achieves good performance rate in detecting CKD via layers. Aditya et al. [13] discovered DL structure-based stacked autoencoder for CKD classification through multimedia data's along with SoftMax classifier. This autoencoder helpful in extracting relevant features from source UCI dataset finally SoftMax classifier used to categorize and predict final class. In this survey, 400 CKD patients along with 25 features were utilized for investigation in which multimodal model achieved 100% classification accuracy in predicting and categorizing CKD. Chin et al. [14] This survey-based on automatic detection of kidney disease via estimated Glomerular Filtration Rate (eGFR) along with status of CKD using DL based transfer learning technique. The dataset utilized is an amalgamation of both ResNet and ImageNet in NN structure to find kidney function depends on 4,505 US kidney images as input. Moreover, NN recognizes the CKD condition clearly defined by an eGFR which is < 60 ml/min.

Himanshu et al. [15] surveyed 224 patients records of CKD accessible in machine learning repository till 2015 to diagnose kidney disease using DL model. This implementation was carried out using cross-validation approach being model secure from over fitting.

Vasanthselvakumar et al. [16] aimed this proposal to identify and categorize CKD mostly kidney stones, cysts in the kidney, and renal failure kidney. Premature forecasting of CKD will accumulate living days from these kinds of inferior diseases. To predict CKD disease earlier, the author projected structure consists of gradient feature along with Adaboost algorithm. Also, Convolutional NN structural design has been used for classifying kidney disease. For predicting CKD, batch prediction model were estimated using US kidney input images.

Charumathi et al. [17] built DL approach to diagnose CKD through retinal images collected data from Singapore epidemiology eye disease (5,188 patients) and Beijing eye disease patients retinal images. The seed validation estimated to diagnose CKD defined by eGFR < 45 mL/minute as in Table 1.1.

TABLE 1.1 Diagnose CKD via SEED Validation

	Metrics (Sensitivity)	
	Image Only	Hybrid Algorithm
SEED validation	93.3%	96%

1.2.1.6   DETECTING CONGENITAL ABNORMALITIES KIDNEY AND URINARY TRACT (CAKUT) DISEASE

Qiang et al. [18] proposed DL-based transfer learning used for feature extraction from kidney Ultrasound images to enhance CAKUT diagnosis in children. Predominantly, a pre-trained ImageNet-Caffe-Alex model is implemented for transfer learning-based feature extraction from 3-features maps worked out from US images, comprised of original images, gradient features, and distanced transform features. Moreover, integration of both transfer learning features along with conventional imaging features produced good performance in differentiating CAKUT from normal patients.

Zheng et al. [19] classified US input images as children with CAKUT disease and normal children using DL approach. The features extracted of kidney disease from US images were done using transfer learning approach by gathered datasets from clinical care of 50 children with abnormal kidney disease also 50 controls. The performance in detecting children kidney disease also classification namely CAKUT estimated using metrics calculation such as specificity, sensitivity, and classification rate shown in Table 1.2.

TABLE 1.2   CAKUT Disease Classification in Specific Region

	Classification Rate	Specificity	Sensitivity
Left kidney	84	84	85
Right kidney	81	74	88
Bilateral abnormal kidney scans	87	88	86

1.2.2   MACHINE LEARNING ALGORITHMS EMPLOYED IN DETECTING KIDNEY DISEASE

1.2.2.1   DIAGNOSE KIDNEY TUMOR DISEASE

Turk et al. [20] In this work, the author provided information regarding diagnosis of kidney cancer which helps the patients who suffer from kidney tumor disease. Several scientists built up various treatments for kidney cancer disease patients; however, this work enhances something better to medical fields, especially for kidney tumor patients. This work mainly intended with finding cancer cells in kidney region through DL as well as machine learning approaches. This method is very useful to convoy patients and also doctors via early diagnosis and classification of disease.

1.2.2.2   PREDICTING CHRONIC KIDNEY DISEASE

Abdullah et al. [21] applied a machine learning algorithm, namely logistic regression, utilized for binary classification, which computes correlation among predictor variable and output variable by calculating likelihood via logistic function. Moreover, classification of kidney disease is possible using machine learning approach. To predict performance in detecting and classifying CKD, metrics such as precision, recall, F1-score, and accuracy were estimated. Anusorn et al. [22] applied machine learning algorithms explicitly Logistic Regression, DT, Support Vector Machine, K-Nearest Neighbor (KNN) to diagnose CKD and classify normal kidney from kidney disease makes use of medicinal fields. Figure 1.3 shows that how this survey identify CKD as well as perform classification to distinguish normal from affected kidney using machine learning approach.

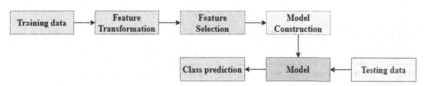

FIGURE 1.3   CKD prediction and classification using machine learning.

Fuzhe et al. [23] In the medical field, patients of CKD increasing day-by-day. The main source of advance treatment for CKD is greater accuracy classification capability using machine learning algorithms which become more significant in medical analysis. This survey suggested Heterogeneous modified ANN for premature findings, segmentation, and analysis of chronic renal malfunction on Internet of Things in medical dais from US kidney images as input. Moreover, this novel method comprises of support vector machine (SVM) and multilayer perceptron

(MLP) with back-propagation algorithm that reaches greater accuracy in finding CKD.

Reem et al. [24] focused on finding CKD using mining and machine learning methods. For predicting CKD based on stages, the dataset accumulated from King Fahd University Hospital (KFUH) in Khobar through Naïve Bayes (NB), KNN, SVM, and Artificial Neural Networks (ANN). The enhanced testing accuracy for SVM, ANN, and NB as 98% whereas KNN reaches accuracy nearly 94%.

Jiongming et al. [25] gathered CKD dataset from UCI machine learning repository contains huge amount of missed parameters. Various machine learning algorithms such as NB, SVM, Logistic Regression (LR), Feedforward NN, Random Forest (RF) and KNN were applied to diagnose medical disease. Also, hybrid model integrated with RF with Perceptron and LR achieves higher accuracy as 99.83%.

Ramya et al. [26] projected several classifiers approaches such as Backpropagation NN, RF, Radial Basis function. The investigator largely paying attention his work that used for reduction in diagnosis time and also to develop the kidney diagnosis accuracy by way of these classification algorithms. Here, Radial basis function algorithm outperforms with better accuracy as 85.3% for classifying the CKD stages.

The stages of CKD can be reviewed in Table 1.3 by Ramya et al. [26].

TABLE 1.3   Stages of CKD for Kidney Disease Classification

Stages	GFR	Description
Stage 1	90 or greater than 90 mL/minute	Kidney damage with normal GFR
Stage 2	60–89 mL/minute	Kidney injure with mild decrease in GFR
Stage 3	30–59 mL/minute	Fair diminish in GFR
Stage 4	15–29 mL/minute	Rigorous drop in GFR
Stage 5	< 15 mL/minute or dialysis	Kidney failure

In this work, the creator has decorated the up-to-date development of essential components of image analysis, as well as data mining (DM) techniques for medical catalog investigation have been utilized to analyze tumors present in kidney. CT scan of kidney images helps how it arises and which stage the cancer in kidney is being diagnosed. Latest trends for kidney tumor recognition and premature prediction techniques have been scrutinized, which shows the necessity of the huge amount of data for training [8]. Also, systematic and meta-analysis have developed into Computer-Aided Diagnosis (CAD) in therapeutic informatics. CAD is necessary in kidney tumor classification and stage prediction. Some methods applied on CT image analysis like preprocessing, segmentation, feature extraction, and finally classification algorithms suitable for distinguishing normal kidney and tumor kidney, which is treated as abnormal.

1.2.2.3 KIDNEY DISEASE PREDICTION USING MINING

Haya et al. [27] forecasted whatever fault happens in kidney function via accomplishment of various DM classifiers approaches such as NB, KNN, DTs, Backpropagation NN, one rule classifiers. Among these DM methods, NB generates enhanced outcome rather than complementary classification methods which has accuracy as 99.3%, sensitivity as 0.97.

Finding renal abnormalities determined by Subarna and Kiran [28] lesions in kidney founded by Zhang et al. [30]; and Chen et al. [38] investigated CKD using machine learning algorithms such as NB, Logistic, RF, AdaBoost, Support Vector Machine. This CKD handling helps in dropping kidney injury progression. Here, while comparison have been done among both models, machine learning algorithms outperform high performance in terms of accuracy.

1.3 PERFORMANCE EVALUATION

1.3.1 ACCURACY ESTIMATION OF VARIOUS SURVEYS VIA DIFFERENT TECHNIQUES

Moreover, in this review, we analyzed many DL techniques for detecting kidney disease by accuracy prediction. Through accuracy calculation, performance in finding kidney disease and classification were scrutinized in Table 1.4 and also in Figure 1.4.

TABLE 1.4 Survey Analysis of Accuracy Prediction Using Deep Learning Techniques

Survey	Objective	Methodology Used	Precision	Recall/ Sensitivity	F1-Score	Accuracy	Outcome
Justin et al. [1]	Finding kidney tumor and non-tumor	How CNN useful in medical applications	–	–	–	–	Classifying input images into tumor and non-tumor
Fuzhe et al. [22]	Premature findings, segmentation, and analysis of chronic renal malfunction	Heterogeneous modified ANN	–	–	–	97.5%	Chronic kidney disease detection using deep learning-based algorithm
Hui Zhang et al. [4]	Detect injury in kidney	RCNN	85.5%	83.8%	83.9%	84%	Finding lesion in kidney images
Hu et al. [5]	Kidney disease classification as normal and malicious	Abnormal behavior detector	–	–	–	79.88%	Detecting abnormal behavior in kidney images
Nour Eldeen et al. [2]	Finding kidney disease, especially tumor	BPSO-DT and CNN	94.96%	95.09%	95.03%	96.9%	Detect tumor through ultrasound kidney images
Turk et al. [20]	Early diagnosis of kidney disease	CNN	–	–	–	–	Classify ultrasound kidney input image as kidney tumor and non-tumor
Vasanth et al. [16]	Categorize chronic kidney disease	CNN	93.7%	93.7%	93.7%	85.2%	Detecting disease and classifying kidney disease

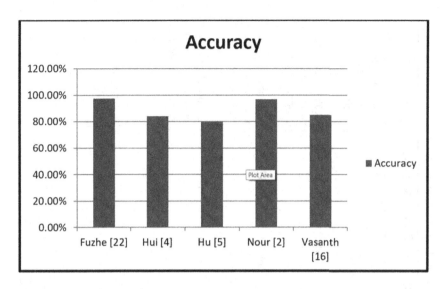

FIGURE 1.4 Accuracy estimation for detecting kidney disease using deep learning techniques on various surveys.

1.3.2 ACCURACY CALCULATION ON DIFFERENT SURVEYS THROUGH MACHINE LEARNING ALGORITHMS

In this literature survey, we investigated some research work in detecting, train model, applying suitable algorithms, validation, and classifying kidney disease as normal and malicious via several machine learning algorithms, which is shown in Table 1.5 and Figure 1.5.

1.3.3 METRICS USED

The model evaluation is carried out through metrics such as sensitivity, Specificity, accuracy too. The formulas used for metrics are described in Table 1.6.

TABLE 1.5 Survey Analysis of Accuracy Prediction in Kidney Disease Using Machine Learning Approaches

Survey	Objective of Work	Methods Used	Sensitivity/Recall	Specificity	Accuracy	Outcome
Fuzhe et al. [22]	Diagnosis of chronic renal failure	• SVM • ANN • MLP	✓	✓	92.3%	Classification of data as benign and malicious
Haya et al. [26]	Predict chronic kidney disease	• Naïve Bayes • KNN • Rule-based classifiers	97%	100%	99.3%	Calculating accuracy to detect CKD via performance estimation
Jiong et al. [24]	Detect chronic kidney disease	• Hybrid model • Logistic regression • Random forest	99.76%	99.93%	99.83%	These methodologies applicable to medical fields in disease diagnosis
Reem et al. [23]	Predict chronic kidney disease	• Naïve Bayes • Artificial neural network • Support vector machine	100%	96.4%	98%	Diagnose chronic kidney disease patients by estimating accuracy
Turk et al. [20]	Classification of kidney disease	• SVM • KNN • Logistic regression	—	—	—	Categorize normal and abnormal

TABLE 1.5 *(Continued)*

Survey	Objective of Work	Methods Used	Sensitivity/Recall	Specificity	Accuracy	Outcome
Theyazn et al. [36]	To enhance the observation detection system for chronic diseases is helpful to save people's life.	• NB • KNN • SVM • RF	89.4% 37% 100% 60.03%	89.4% 40% 96% 74.4%	95.9% 50.2% 98.07% 69.57%	Enhance accuracy of classification algorithms in detecting CKD
Setu Basak et al. [34]	Predicting and staging CKD of diabetes disease	• NB • Instance-based learning • RF • Decision tree (J48) • Decision stump	95% 98% 87% 97% 97%	— 	95.3% 98.3% 88% 97% 97.8%	Classify CKD patients with diabetes and normal people from gathered data source

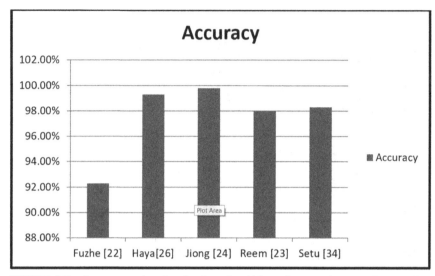

FIGURE 1.5   Accuracy estimation for detecting kidney disease using machine learning algorithms on various investigations.

TABLE 1.6   Metrics Formula

Metrics	Described Formulas
Sensitivity/recall	
Specificity	
Accuracy	

1.3.4   STATE OF ART IN THE FIELD OF DISEASE PREDICTION (SPECIFICALLY KIDNEY)

Table 1.7 demonstrates how DL approaches helpful in medicinal domains in diagnosing disease through source data, gathering input data, what kind of issues occur, apply appropriate DL algorithm to resolve the issues, and finally what the production will be. This production is really helpful in medical personnel such as doctors, nurses, etc.

TABLE 1.7　State-of-Art in Disease (Kidney) Prediction Domain

Research Title	Input	Deep Learning Model	Application Domain	Source Data	Production
Prediction of kidney function [27]	Biopsy images	CNN; hyperparameter tuning	Chronic kidney disease	NEPTUNE dataset; Kidney slide resolution (150,000 × 50,000 × 3) pixels	Predict eGFR absolute error as 30.5
Diagnosis of chronic kidney disease [22]	Ultrasound images	Heterogeneous modified artificial neural network (MLP, BP)	Chronic kidney disease	CKD from UCI machine learning repository	Accuracy prediction as 96.5%
Risk level prediction of CKD [31]	Attribute relation file format	ANN with fuzzy based rule algorithm	Chronic kidney disease at risk stage	CKD from WEKA dataset	Neuro-fuzzy accuracy as 97%
Prediction of CKD [29]	CT images	Adaptive hybridized deep convolutional neural network	Early diagnosis and prediction of CKD	http://www.mediafire.com/datasets.	Finding classification accuracy in predicting CKD as 97.3%
Computer-aided detection [4]	2 D CT scan images	Convolutional neural network	Evaluate CNN performance on CAD applications as well as lymph node detection	ILD dataset	Classification accuracy rate as 94%
CKD prediction [35]	CT angiography	Multiple logistic regression	Evaluation of CKD by eGFR rate	105 patients data	CKD diagnosis from patients CT angiography as 89.4% accuracy

TABLE 1.7 *(Continued)*

Research Title	Input	Deep Learning Model	Application Domain	Source Data	Production
Diagnosis of diabetes [33]	Based on patients parameters	Deep neural network (5-fold, 10 fold validation)	Predicting disease especially diabetes patients	Pima Indian diabetes data from UCI machine learning repository	Prediction accuracy of diabetes 98.35%
Predicting kidney disease patients with hypertension [37]	Patients records	Hybrid neural network (BiLSTM, autoencoder)	Predict patient with hypertension	Amount of EHR data, 12 cities in China	Accuracy in kidney disease diagnosis with hypertension as 89.7%
CKD prediction [32]	Ultra-sonography images	Glomerular filtration rate	Diagnosing chronic kidney disease	120 patients with stage 1–5 CKD	Correlation among e-GFR and ultrasound parameters

1.4 CONCLUSION

In current years, several machine learning and DL techniques have been utilized for analysis as well as detection of diseases such as kidney, heart, lungs, and every part of the human body that makes helpful in medicinal domains. Every issue needs an understanding of issues and makes a diagnosis the very most suitable machine learning and DL algorithms which are frequently indiscriminate analysis having several layers and features too. In this work, we surveyed machine learning and DL algorithms suitable for predicting kidney disease and classifying disease via US images as input. Finally, accuracy comparison of different surveys makes how far these techniques applicable in kidney disease diagnosis and also classifying kidney disease as normal from abnormal (deviations in kidney like tumor arises, renal failure, etc.).

KEYWORDS

- artificial neural networks
- chronic kidney disease
- computed tomography
- congenital abnormalities kidney and urinary tract
- convolutional neural network
- positron emission tomography

REFERENCES

1. Justin, K., Wang, L., Rao, J., & Lim, T., (2018). Deep learning applications in medical image analysis. In: *IEEE Access* (Vol. 6, pp. 9375–9389).
2. Nour, E. M. K., Taha, M. H. N., Ezzat, A. D., Slowik, A., & Hassanien, A. E., (2020). Artificial intelligence technique for gene expression by tumor RNA-seq data: A novel optimized deep learning approach. *IEEE Access, 8*, 22874–22883.
3. Cao, C., Liu, F., Tan, H., Song, D., Shu, W., Li, W., Zhou, Y., et al., (2018). Deep learning and its applications in biomedicine. *Genomics, Proteomics & Bioinformatics, 16*(1), 17–32.

4. Hoo-Chang, S., Holger, R. R., Mingchen, G., Le Lu, Sr., Ziyue, X., Isabella, N., Jianhua, Y., et al., (2016). Summers deep convolutional neural networks for computer-aided detection: CNN architectures, dataset characteristics and transfer learning. *IEEE Transactions on Medical Imaging, 35*(5).

5. Hu, Y., (2020). Design and implementation of abnormal behavior detection based on deep intelligent analysis algorithms in massive video surveillance. *J. Grid Computing, 18*, 227–237.

6. Domingues, I., Pereira, G., Martins, P., et al., (2020). Using deep learning techniques in medical imaging: A systematic review of applications on CT and PET. *Artif. Intell. Rev., 53*, 4093–4160.

7. Vinayagam, P., Sreemathi, M., Jeevitha, K., & Sandhya, S., (2019). Kidney stone detection using neural network. *International Journal of Applied Engineering Research, 14*(6). ISSN 0973-4562.

8. Guang-Bin, H., (2003). Learning capability and storage capacity of two hidden-layer feed-forward networks. *IEEE Transactions on Neural Networks, 14*(2), 274–281.

9. Jackson, P., Hardcastle, N., Dawe, N., Kron, T., Hofman, M. S., & Hicks, R. J., (2018). Deep learning renal segmentation for fully automated radiation dose estimation in unsealed source therapy. *Frontiers in Oncology, 8*, 215.

10. Timothy, L. K., Panagiotis, K., Marie, E. E., Jaime, D. B., Frank, S. C., Peter, C. H., Bernard, F. K., et al., (2017). Performance of an artificial multi-observer deep neural network for fully automated segmentation of polycystic kidneys. *J. Digit Imaging, 30*, 442–448.

11. Vijaya, B. K., Priyamvada, S., Christopher, Q. L., Dan, M., Mostafa, E. B., Joel, M. H., Jean, M. F., et al., (2018). Association of pathological fibrosis with renal survival using deep neural networks. *Kidney International Reports, 3*(2), 464–475.

12. Abdullah Al, I., Md Nur, A., & Fatema, T. J., (2018). Classification of chronic kidney disease using logistic regression, feedforward neural network and wide & deep learning. *International Conference on Innovation in Engineering and Technology (ICIET)* (pp. 27–29). ISBN: 978-1-5386-5230-5.

13. Aditya, K., Gurinder, S., Babita, P., Shrasti, T., Deepak, G., & Ashish, K., (2019). KDSAE: Chronic kidney disease classification with multimedia data learning using deep-stacked autoencoder network. *Multimedia Tools and Applications* (pp. 1–16). Springer Nature.

14. Chin-Chi, K., Chun-Min, C., Kuan-Ting, L., Wei-Kai, L., Hsiu-Yin, C., Chih-Wei, C., Meng-Ru, H., et al., (2019). Automation of the kidney function prediction and classification through ultrasound-based kidney imaging using deep learning. *Npj Digital Medicine, 2*(1).

15. Kriplani, H., Patel, B., & Roy, S., (2019). Prediction of chronic kidney diseases using deep artificial neural network technique. *Computer-Aided Intervention and Diagnostics in Clinical and Medical Images* (pp. 179–187). Springer, Cham. Chicago.

16. Vasanthselvakumar, R., Balasubramanian, M., & Palanivel, S., (2019). Detection and classification of kidney disorders using deep learning method. *J. Mech. Cont.& Math. Sci., 14*(2), 258–270.

17. Charumathi, S., Dejiang, X., Daniel, S. W. T., Simon, N., Riswana, B., Haslina, H., Cynthia, L., et al., (2020). A deep learning algorithm to detect chronic kidney disease

from retinal photographs in community-based populations. *The Lancet Digital-Health, 2*, 1–8.

18. Qiang, Z., Gregory, T., & Yong, F., (2018). Transfer learning for diagnosis of congenital abnormalities of the kidney and urinary tract in children based on ultrasound imaging data. *Proc. IEEE Int. Symp. Biomed. Imaging*, 1487–1490.

19. Zheng, Q., Furth, S. L., Tasian, G. E., & Fan, Y., (2019). Computer-aided diagnosis of congenital abnormalities of the kidney and urinary tract in children based on ultrasound imaging data by integrating texture image features and deep transfer learning image features. *Journal of Pediatric Urology, 15*(1), 75.e1–75.e7.

20. Fuat, T., Murat, L., & Necaattin, B., (2019). Machine learning of kidney tumors and diagnosis and classification by deep learning methods. *International Journal of Engineering Research and Development, UMAGD, 11*(3), 802–812.

21. Anusorn, C., Thipwan, F., Tippawan, N., Wandee, C., Sathit, S., & Nitat, N., (2016). Predictive analytics for chronic kidney disease using machine learning techniques. *The 2016 Management and Innovation Technology International Conference (MITiCON-2016)* (pp. 80–83).

22. Ma, F., Sun, T., Liu, L., & Jing, H., (2020). "Detection and diagnosis of chronic kidney disease using deep learning-based heterogeneous modified artificial neural network. *Future Gener. Comput. Syst., 111*, 17–26.

23. Reem, A. A., *et al.,* (2018). Preemptive diagnosis of chronic kidney disease using machine learning techniques. In: *2018 International Conference on Innovations in Information Technology (IIT)* (pp. 99–104).

24. Qin, J., Chen, L., Liu, Y., Liu, C., Feng, C., & Chen, B., (2020). A machine learning methodology for diagnosing chronic kidney disease. In: *IEEE Access* (Vol. 8, pp. 20991–21002).

25. Ramya, S., & Radha, N., (2016). Diagnosis of chronic kidney disease using machine learning algorithms. *International Journal of Innovative Research in Computer and Communication Engineering, 4*(1).

26. Alasker, H., Alharkan, S., Alharkan, W., Zaki, A., & Riza, L. S., (2017). Detection of kidney disease using various intelligent classifiers. In: *2017 3rd International Conference on Science in Information Technology (ICSITech)* (pp. 681–684).

27. Ledbetter, D., Ho, L., & Lemley, K. V., (2017). *Prediction of Kidney Function from Biopsy Images Using Convolutional Neural Networks* (pp. 1–11). Los Alamos National Lab: Santa Fe, NM, USA.

28. Subarna, C., & Kiran, R. P., (2020). Diagnosis of kidney renal cell tumor through clinical data mining and CT scan image processing: A survey. *International Journal of Research in Pharmaceutical Sciences, 11*(1), 13–24.

29. Chen, G., et al., (2020). Prediction of chronic kidney disease using adaptive hybridized deep convolutional neural network on the internet of medical things platform. In: *IEEE Access, 8*, 100497–100508.

30. Zhang, H., Chen, Y., Song, Y., Xiong, Z., Yang, Y., & Jonathan Wu, Q. M., (2019). Automatic kidney lesion detection for Ct images using morphological cascaded convolutional neural networks. In: *IEEE Access* (Vol. 7, pp. 83001–83011).

31. Kerina, B. C., Noorie, H., Ronnie, C. D., & Ch Iyengar, S. N. N., (2017). Risk level prediction of chronic kidney disease using neuro-fuzzy and hierarchical clustering

algorithm (s). *International Journal of Multimedia and Ubiquitous Engineering,* *12*(8), 23–36.

32. Mustafa, Y., Özgür, Ç., Mehmet, N. T., Ramazan, D., Selçuk, A., Elif, D., Mustafa, Y., & Faruk, T., (2017). Role of ultrasonographic chronic kidney disease score in the assessment of chronic kidney disease. *Int Urol. Nephrol., 49*(1), 123–131.

33. Safial, I. A., & Md Milon, I., (2019). Diabetes prediction: A deep learning approach. *I. J. Information Engineering and Electronic Business, 11*(2), 21–27.

34. Setu, B., Md Mahbub, A., Aniruddha, R., Ahmed Al, M., & Anup, M., (2019). Predicting and staging chronic kidney disease of diabetes (type-2) patient using machine learning algorithms. *International Journal of Innovative Technology and Exploring Engineering (IJITEE)* (Vol. 8, No. 12). ISSN: 2278-3075.

35. Sung-Hye, Y., Deuk, J. S., Kyung-Sook, Y., Myung-Gyu, K., Na Yeon, H., Beom, J. P., & Min Ju, K., (2019). Predicting the development of surgically induced chronic kidney disease after total nephrectomy using body surface area-adjusted renal cortical volume on CT angiography. American Journal of Roentgenology, *212*(2), W32–W40.

36. Theyazn, H. H A., Ali, S. A., & Mohammed, Y. A., (2020). Soft clustering for enhancing the diagnosis of chronic diseases over machine learning algorithms. *Journal of Healthcare Engineering, 2020*(4984967), 1–16, https://doi.org/10.1155/2020/4984967.

37. Yafeng, R., Hao, F., Xiaohui, L., Donghong, J., & Ming, C., (2019). A hybrid neural network model for predicting kidney disease in hypertension patients based on electronic health records. *BMC Medical Informatics and Decision Making, 19*(51).

38. Chen, Z., Zhang, X., & Zhang, Z., (2016). Clinical risk assessment of patients with chronic kidney disease by using clinical data and multivariate models. *Int. Urol. Nephrol., 48*(12), 2069–2075.

CHAPTER 2

Deep Learning-Based Computer-Aided Diagnosis System

G. VIJAYA

Department of Computer Science and Engineering,
Bapatla Women's Engineering College, Bapatla, Andhra Pradesh, India,
E-mail: viji.pooshan@gmail.com

ABSTRACT

Based on the fact that computing technology rises unconditionally, alive machine learning algorithms have proven knowledge to implement complex imaging modalities. Studies are fetched out from the past few decades to nurture the computer-aided diagnosis (CAD) system to give a hand to radiologist for connecting diverse actions of diseases. This system's decisive purpose is to comfort the physician to limelight the anomalous findings of the imaging modalities by dropping the interpretation time. Timely diagnosis of the disease permits the physicians to afford the right treatment and expand their continual existence. On the other hand, the physical diagnosis is often arduous because of the outstanding number of imaging modalities and can be highly biased inter-observer patchiness. Swift progress in machine learning guides the CAD systems in an innovative way using deep learning (DL) skills, which can extend the performance up to human levels. Besides, this chapter figured out the opportunities and obstacles for designing and learning competent, operative, and vigorous deep learning algorithms in medical analysis and derives attention to the direction of forthcoming investigations.

Application of Deep Learning Methods in Healthcare and Medical Science.
Rohit Tanwar, PhD, Prashant Kumar, PhD, Malay Kumar, PhD, & Neha Nandal, PhD (Editors)
© 2023 Apple Academic Press, Inc. Co-published with CRC Press (Taylor & Francis)

2.1 INTRODUCTION

Medical imaging (MI) is a step-by-step process to visualize the internal organs of a body for therapeutic analysis and ocular illustration of the function of organs/tissues. MI is ought to expose the body's interior structure, which was covered by skin and bones, and it is also useful to spot out the abnormalities for further treatment of the diseases. MI is not only helping the physician to identify the disease but is also helpful to create a standard database for future reference. Currently, MI plays a crucial role in exposing the information about various diseases as like [1–3] and these procedures normally involve skilled doctors to interpret it. Besides, for the same patient itself, based on the diagnosis, the doctor may take an altered decision at a different time. From Figure 2.1 different imaging modalities along with the duration and diagnoses are explained.

To attain an added consistency and precise analysis, recently, in the MI field, computer-aided detection and diagnosis (CADe and CADx), are structures that support physicians by analyzing the health representation process such as X-ray, tomographic scans such as computed tomography (CT), magnetic resonance imaging (MRI), positron emission tomography (PET), and ultrasound (US) images, which in turn, gives a great deal of information within a short-sperm of time. CADe systems are typically curbed to highlight the noticeable structures and segments while CADx systems assess the noticeable structures. Computer-aided diagnosis (CAD) is the central field of medical research since the invention of X-rays [4–13]. After then, the physicians could see inside the patients' body without dissection. As X-ray is the first medical image, it does not work well in soft tissues, which in turn makes the development of 'US images.' In the late 20th century, after the invention of tomography images like CT and PET becomes more popular for producing three-dimensional (3-D) images. It is tough to exaggerate how much the field of ML will undergo changes within a single century. With the existing technologies, nowadays, it is possible to identify the disease without any human intervention. Figure 2.2 gives a step-by-step procedure for a generic CADe and CADx structure of segmenting and classification of lesions in medicinal descriptions.

During the 21st century, deep learning (DL) – a branch of artificial intelligence (AI) raises the curtain of the 'medical industry' because of its capability to hold a huge amount of features in an unstructured data. For instance, an RGB image of size 800 × 1,000 pixels has 2.4 features, which is too far to handle by the traditional machine learning algorithms. Hence, DL becomes more popular not only for handling the data but it also has supreme accuracy with exceptional speed. DL will drive the healthcare industry in a different path, and according to my perception, in the forthcoming years, there must be an increase in the number of influences of DL in both the academic and industry-oriented efforts.

Image modalities	Time taken	Imaging method	Common diagnosis
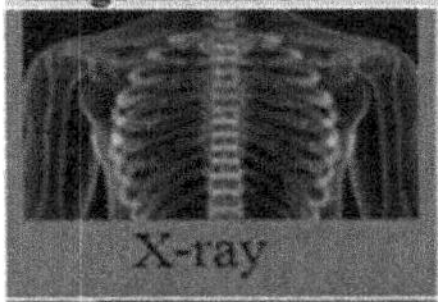 X-ray	10 to 15 minutes	Ionizing radiation	❖ Arthritis ❖ Bone fracture ❖ Digestive tract problems ❖ Infections
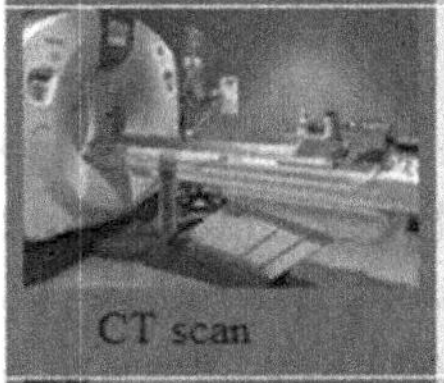 CT scan	10 to 15 minutes	Ionizing radiation	❖ Tumors & cancers ❖ Heart disease ❖ Bone fractures ❖ Vascular disease ❖ Guide biopsies
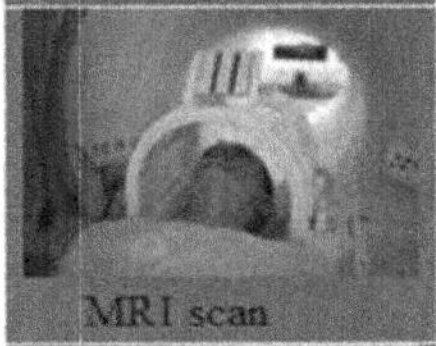 MRI scan	45 minutes to 1 hour	Magnetic waves	❖ Stroke ❖ Blood vessel issues ❖ Spinal cord disorders ❖ Joint or tendon injuries
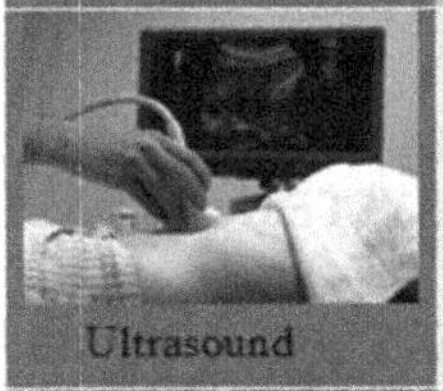 Ultrasound	30 minutes to 1 hour	Sound waves	❖ Monitoring pregnancy ❖ Breast lumps ❖ Joint inflammation ❖ Genital/prostate issues ❖ Gallbladder disease
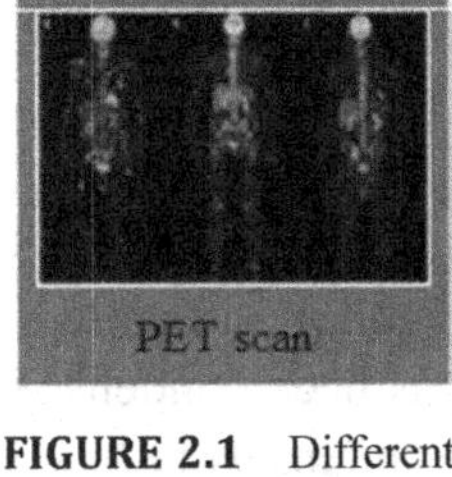 PET scan	$1\frac{1}{2}$ to 2 hours	Radio tracers	❖ Alzheimer's disease ❖ Parkinson's disease ❖ Coronary artery disease ❖ Epilepsy ❖ Seizures

FIGURE 2.1 Different types of image modalities, time taken to retrieve the image and the common diagnosis.

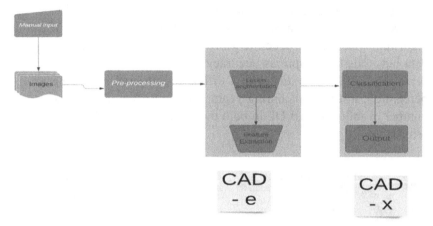

FIGURE 2.2   Flowchart for a generic CADe and CADx system.

Hence, this chapter is categorized as: Section 2.2 expresses the history of the development of CAD systems, starting from the X-rays to real-time imaging systems. Section 2.3 gives you a trailer of future enhancement of imaging technologies. Sections 2.4 and 2.5 stretches out the basics of DL and the overview of DL in the healthcare field. In Section 2.6, a novel prototype for CAD system was proposed. It is not the only choice to implement the CAD system, but it is a basic model to detect and diagnose the disease using computerized algorithms. Section 2.7 expresses its views relating to the ethical implications and the future inferences of the DL-based CAD systems.

2.2   CHRONOLOGICAL DEVELOPMENT OF CAD SYSTEM

Before the arrival of MI, the doctors' roles are critical as they diagnosed the diseases without seeing their "insides," apart from surgery. Then and there, the invention of the X-ray in 1865, noteworthy changes have been in MI because X-rays are at a sensible price and readily accessible. Moreover, customary X-rays are not suitable for fine lesions and CT images had replaced them. CT images themselves have gone through modifications throughout the years. The dimension of image piece has been condensed as soon as the helical CT comes on the scene – which has intensely reduced the image acquisition time.

MRI is a detailed imaging tool that produces 3-D whole images which is commonly used to highlight the ailment, analyzing, and monitoring it for further treatment. MRI had been materialized at the beginning of the 21st century, where MI underwent radiation exposure that remained at a peak. This imaging system's procedure is that it uses magnetic fields to obtain images of soft tissues or non-bony body structures. On the other hand, for more detailed imaging, contrast agents are injected into the patient introrsely to get crystal images.

Ultrasound is a form of imaging in the medical field used to identify any changes in size, shape, or outline of tissues, and organs or to spot anomalous masses. Ultrasound uses greater-frequency radio waves which are penetrated to the core of the body and in turn, it produces echoes. Through these echoes, it is promising to conclude how distant the object is and calculate the shape, size, and consistency of the objects. As soon as these echoes struck the transducer, they cause electrical signals which were transmitted to the US scanner. With the back-and-forth process of each echo, the scanner computes the distance of the tissue boundary. Based on these distances two dimensional (2D) images of tissues/organs are generated. But it is inactive in soft tissues or any organs that have air, like the lungs. One of the exceptional customs of US is to observe the fetus's growth for the complete period of pregnancy, together with the imaging of all organs like the heart, muscles, and blood vessels. For the period of examination, the physicians apply lotion to the skin. These preserve air pockets to be produced among the skin and the transducer that prevents US waves from passing into the body. Ultrasound diagnosis is generally considered safe and does not create ionizing radiation just like X-rays. Moreover, US is capable of making some side-effects in the physique under particular situations and circumstances (Table 2.1).

Additional to the existing MI practices, PET emits radiolabeled compounds, for instance, dextrose which is perceived through body cells. When it is noticed by an external stimulus, the divisions of these compounds provide a hint to the opinion. The inhalation of contrast dye has spot-out the specific site through imaging. This aids in identifying vascular abnormalities and bleeders. PET scans can also estimate oxygen consumption, blood flow, sugar usage in the body, and metabolism of organs at the cellular level. This is vital because ailment often originates at the cellular level. CT and MRI scans can't disclose problems at this earlier stage. Usually, all the current imaging methods are pooled to give the medical practitioners a distinct plan to know exactly about the patient's condition.

TABLE 2.1   Various Imaging Modalities with Remarks

Authors	Imaging Modality	Dataset	Remarks
Koitka et al. [14]	X-ray	RSNA pediatric bone age challenge	Inception ResNetV2 as feature extractor and DL to inspect the age of fossil
Deniz et al. [15]	MRI	Volumetric structural MR image of the proximal femur	UNets to segment proximal–femur–RoI for fracture risk assessment
Abd-Ellah et al. [16]	MRI	Reference image database to evaluate response (RIDER) neuro MRI database	CNN with error-correcting support vector machine (ECSVM) for tumor detection and R-CNN for tumor localization
Konstantinos et al. [17]	MRI	BRATS2015 AND ISLES 2015	3D CNN to segment lesions from brain images with ischemic stroke

2.2.1   EVOLUTION OF MEDICAL IMAGING TECHNOLOGY

MI technology had undergone unexpectedly rapid progress over the centuries and the demand for MI has bloomed at a freakish rate. The imaging modalities are not stick-on to a specific application due to increase in sophistication on further processing and varieties of more progressive ways for loading and keeping medical images. The aim here is to mine the supreme profit from the recent technologies and extend it to the maximum amount of individuals possible. In therapeutic imaging currently, physicians can hold images with exceptional perceptions and gather details from a related set of values.

2.2.1.1   PROGRESSES IN REPOSITORY AND RECOVERY OF VISUALIZING DATA

Using discrete forms of visualizing procedures hired currently and the typical details that they give, amalgamation, and easiness of cooperation are of supreme concern to medication industries and consumers. More or less all kinds of real-world visualizations are computerized and it comprises massive data sets. The subsequent launch of Picture Archiving and Communications System (PACS) is a forum for consolidated repository and retrieval of diagnostic images from diverse plans. Images are generally warehoused in the Digital Imaging and Communications in Medicine (DICOM) pattern in the PACS server.

2.2.1.2 INNOVATIVE MEDICINAL IMAGING DEVICES

1. **3D Therapeutic Imaging Mechanization:** A solitary negative impact of current medicinal visualizing procedures is represented as two-dimensional images even though body muscles and organs are 3-D. Because, framing a 3D structure from the existing one, physicians have to alter the image portions at different angles and then reformulating it to give a spontaneous image for analysis. This leads to human errors and also a time-consuming practice. 3D imaging is among today's innovative MI devices having higher resolution with fewer artifacts. The rapid acquirement of images using fine slices and enhanced rendering algorithms may speed up to yield amazing 3D images in predefined formats. Moreover, based on the assessment of the live Multi-detector Computed Tomography (MDCT) and MR scanners, they frequently produce hundreds or thousands of slices per the study, gives noteworthy challenges to clinicians' efficiency. For diagnosis and medical test, the accessibility of 3D images grants radiologists to review the whole anatomy in a few courses and then make it as an analogy for future observation and confirmation with the original 2D images. As a result, 3D restoration is more often a valuable procedure to summarize the vast number of slices crisply and clearly produced by the current MR and CT examiners. Within the domain of radiology, 3D imaging along with virtual reality (VR) takes the scanning technologies step further, permitting clinicians to see the whole part of the body convincingly.

2. **Nuclear/Internal Imaging:** Diagnostician practice internal imaging to rapidly detect interior complications through radioactive materials where the concentration is more in the body.

Normally, in examinations like single-photon emission computed tomography (SPECT) or PET, the clinicians shoot a small quantity of radioactive substance to patients which are observed by the tissues having high blood circulation, which in turn used to focus the possible issues in that organ. For instance, if a person is affected by Alzheimer's disease (AD) his brain contains a substance called amyloid plaques was produced, which can be easily recognized through the existing nuclear imaging. Conventionally, these signs could only be confidently known in an autopsy course, so this imaging will alarm the patient and his family members before the disease gets worse.

3. **Intraoral Imaging:** Throughout surgical procedure, physicians may hire visualizing tools to monitor their surgeon. Intraoral imaging will assist surgeons to differentiate typical and atypical organs, such as lump nodes, and can aid them in making indispensable slices and dissections for the entire operation period.

   In older days, real-time imaging is not on the scene for surgical purposes—because of the reason that the little amount of diffusion necessary for the scan might have an effect on the health of the radiologists. Today's wireless technology similar to the mobile computed tomography (MCT) scanner takes enhanced quality deprived and reveals higher diffusion mutually with the clinicians and also the subject. The CT scanner can also monitor the effect of post-surgery and confirms that there is no internal hemorrhage or any complications related to surgery.

4. **Wearable Technology:** In today's digitalized world, numerous mobile medicinal tools is used to spot out or gather information regarding patients' health from the remote location. By means of this equipment's physicians can have a better understanding of one's health issues without physically touching the patients and can have a direct visualization of the impact on it. The simplest and most unique form of wearable technology/wearable fitness hunters is wristbands fitted with sensors to track their activities and heart rate. They integrate wearers with fitness and health references by synchronizing to numerous smartphone applications.

   One of the popular versions of wearable technology is Wrist bands and watch as shown in Figure 2.3. Consumers were fascinated by its silky look and capability to track their movement all over the day with the device's 5 display lights. In older days watches were only used to indicate time, smartwatches have nowadays been renovated into clinically feasible healthcare tools. The main usage of wearable technology is: it is low-cost, handy, and it clears the way to detect a problem without any medicinal settings and fine-tuning the medications. Significant complications could be pointed out without difficulty, and the victim could be keeping on touch for further attention if needed.

FIGURE 2.3   Smart health watch [18].

5. **Image Fusion:** A progressive medicinal visualizing tool called image fusion is represented in numerous DICOM approaches. Incorporating the information enclosed in numerous images of the identical picture into one amalgamated picture, picture element image fusion is recognized as one of the big implications together with MI. The input is the origin image which is grabbed from diverse imaging tools or a specific detector under distinct variable settings. The combined image should be convenient for an individual or machine observation than any single input. The much more widely used image fusion approaches are PET with MRI and PET with CT scan images, which takes the advantages of each scan and the resultant image must be a precise one. PET lends a hand to identify and confined the area of concern (typically a harmful or infected area), while CT scan highlights outstanding body details such as lesion identification and the nerves associated

with it. The main usage of MRI scan is to draw soft tissue with high-quality. When it is merged there is an amazing growth in the subtlety and relevance of image diagnosis. As in Figure 2.4, image fusion technologies were carried out through MRI, PET with discrete wavelet transform (DWT) and fused rule.

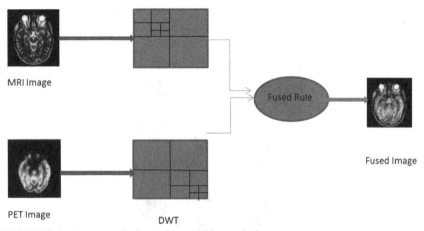

FIGURE 2.4   An example for multi-modal image fusion.

6. **Concurrent Imaging:** Generally, it is a fact that there must be a 'fall-off' in connecting with the time taken by the image which is received and clarified. The fall-off is the junction where the system used to practice and make the image to display it to the health supporters, and in turn they observe each part of the visualization by applying their skill to consolidate it. This fall-off gives notable influence on medicinal end result, particularly in calamities say trauma, at which time is the important impact on it. Nowadays, numerous picturing techniques propose 'concurrent' outcomes, which imply that the space among image gaining and analysis is at least minimum or null.

Despite the fact that the patient is in the imaging unit, medical specialists can monitor the live images on the screen. This not only reduces the fall-off but also has an added advantage of seeing the function of the body at work and therefore evaluating their practical ethics. For instance, dysphasia, the difficulty of swallowing can be estimated using the real-time monitoring of the esophagus or throat. Correspondingly, fetal activities can be perceived from the real-time images using sonography. The influences of concurrent imaging formulate it to be within reach of physicians for making choices intraoperative.

2.3 A PREVIEW OF FUTURE MEDICAL IMAGING TECHNOLOGY

2.3.1 ARTIFICIAL INTELLIGENCE (AI)

AI talks about the feasibility of devices racially mixed with the human brain. This predominantly relates to cognitive tasks, like training and appealing knowledge. The main purpose of AI in medicinal imaging is to plug freaks in human structures, by assisting the physicians to identify the diseases and nursing their therapy. Developments in computational power and massive data sets of patients' information, medical AI can filter this information at extraordinary speeds. This in turn initiates progress. Similarly, AI is learned from the existing algorithm to highlight the anomalies which are not visible to our bared eyes in order to increase the testing precision. Moreover, the subjects' electronic medical record (EMR) is matched with the latest scans, and the anomalies or any malfunctions have to be pinpoint using AI. Added features collected from the subjects' EMR, like any relevant case study can be recovered and used for verification. In the interim, imaging-based algorithms can be drawn on as a calibrating technique for double-checking the diagnoses by physicians without compromising the accuracy. Moreover, the program makers are not the one that makes use of the system in the clinical study. As a result, the clinicians have to have some knowledge about the programs that they want and the programs have some knowledge about the existing clinical terms.

2.3.2 CLOUD-BASED APPLICATIONS (CBA)

Along with the express growth in imaging tools and the global use of medicinal images in preventive medicine have stemmed a resolution to find state-of-the-art methods to store and share data related to MI. The prolonged struggle of storage capacities of millions and billions of patients' information in their local servers has now come to an end after the arrival of Cloud storage. Transferring data to the cloud has mitigated the concerns related to storage and proposes a lesser amount of software and hardware support and eradicating system-oriented problems in local servers. Cloud computing environment allows the storage allocation of data irrespective of the topographical area by means of the web. CBA related to medicinal imaging mitigates the storehouse and regaining of imaging files in the DICOM pattern. Resources gathered such as medical images, processing, and retrieval are assembled by the supplier to several users from a principal location according to each user's constraint. CBA also employs blockchain process – a method of adding a fresh digital document to an existing pattern, such as the addition of a new connection to an already existing one. Images which have existence in the cloud storage can be further attached to a blockchain; then and there medical information of a patient is easily reached to any doctor everywhere around the world.

2.3.3   *POST-DICOM – A GROUNDBREAKING OF MEDICAL IMAGING TECHNOLOGY*

Post-DICOM pools the pre-eminent medicinal imaging tool in the futuristic world. The online rendering services enables the PACS system whose medical records storage and retrievals are provided by the medical-based cloud server. DICOM is the widespread transmission etiquette and file format meant for PACS servers. While uploading to Post-DICOM, the total images gained from health imaging modalities (such as CT, MRI, PET, and US) and associated documents (in BMP, JPG, PDF, and AVI formats) are transformed into the DICOM format and warehoused anywhere in the archive while they will be indefinitely staying in it as long as the users' account is active.

Post-DICOM can be used by physicians, radiologists, hospitals, and patients as a suitable and available means to speed up the communication and discussion between these groups and subsequently to increase the problem-solving process and refine patient care.

2.4   A GLIMPSE OF DEEP LEARNING

Even though AI is one of the distinguished technology terms in the Healthcare industry, DL is a division of AI that provides a rich look to MI technology. Contradictory to customary machine learning algorithms DL is powered by huge amounts of data, and it involves high-end mechanisms with commanding GPUs to run within the time limit. Most of the existing machine learning techniques need an expert opinion to cut down the data's intricacy and create patterns orderly to execute it directly. The principal benefit of using DL is that it pops up the raw data features without any human intervention, which has to be depicted in Figure 2.5. DL and the therapeutic industry can traverse data at exemption speeds without conceding the accuracy. In multifaceted problems whenever a greater level of automation is necessary and there is a shortage of domain knowledge for feature engineering (e.g., Robotics, Natural Language Processing, etc.), DL techniques are rise steeply by attaining higher levels of accuracy.

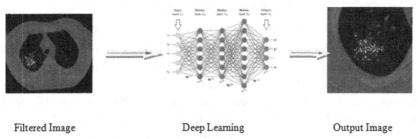

Filtered Image   Deep Learning   Output Image

FIGURE 2.5 Deep learning techniques.

In the field of radiology, DL detects multifaceted patterns spontaneously and relief's radiologists' burden to make intellectual choices reviewing images such as CT, MRI, PET, and radiology reports. The DL-based spontaneous detection and diagnosis performance is almost the same as that of the skilled radiologist. Moreover, customary ML algorithms are inefficient to solve the complications of fitness-related problems having complications and seriousness of the topic. A study [19] shows that nearly 10% of the deaths are due to diagnostic errors. DL is one of the remodeling techniques to cut down these errors and diminish the death rate.

From the moment when it's developed, AI has been over-anxious but poor in delivering.

- Paramount to the till date customary ML approach in machine vision, it depends densely on patterns which are governed by the expert people of the particular field.

- The process is an occupational one, as data differs from person to person and the acquisition of data shows a discrepancy with the knowledge of the experts. Hence, customary learning techniques are not reliable.
- Some of the customary ML algorithms such as Decision Trees (DTs), Support Vector Machines (SVMs), K-Nearest Neighbors (KNNs), Logistic Regression, etc., which are supposed to take raw data for interpretation without the knowledge of hidden representations.
- Besides, before feed the raw data, it has undergone preprocessing based on the medical professional's information and it's very time-consuming.
- Even though the most common image acquisition devices like X-ray, MRI, and CT scans have enriched after a while with high quality, automatic imaging is the toughest one to reach.

The efficacy of DL in healthcare has revealed substantial advancement in seizing invisible patterns and draw out the traits from the algorithms. The applicability of these feature extraction without compromising the accuracy, helps the physician in diagnoses. Some of the existing DL architectures like Convolutional Neural Networks (CNNs), Recurrent Neural Networks (RNNs), deep convolutional neural networks (DNN) and a combination of DL along with some efficient ML algorithms like Support Vector Machine (SVM) also bump into the field. Based on the literature review [20–23], supervised DL techniques are outperformed than unsupervised algorithms related to medical image processing.

2.5   DEEP LEARNING FOR HEALTHCARE IN A NUTSHELL

The central role of image diagnosis is to pinpoint irregularities. DL methods assist the physician by analyzing information and addressing misdiagnosis skillfully and forecasting the consequence of the procedure. There are some well-known DL methods in the medical management occupation that they give results to issues stretches from diagnosing the illness to personalized advice.

The integration of the latest progressive methods, enhanced processing power, and budding interest in state-of-the-art methods of forecasting with low-cost healthcare makes DL a paramount method. A number of MI applications succeeded through DL are discussed in subsections.

2.5.1 DIABETIC RETINOPATHY (DR)

DR is an eye complication of diabetics,' resulting in everlasting blindness based on patients' diabetic stage severity. For screening this disease, the widely used image is the retina owing to its hypersensitive nature. The estimation of the acuteness and the level of sight-threatening interconnected with diabetics are now carried out by experts based on the patients' retinal or fundus images. According to World Health Organization (WHO), the quantity of patients who have diabetes is growing exponentially, which in turn will increase the integer of retinal images obtained through screening. Conversely, it hosts overburden to the clinicians and successively makes a price-hike to the healthcare amenities. This could be relieved using the computer-based automatic system as a supporting tool for medical experts or a complete diagnosis tool.

A clinically feasible automated system can sort out the retinal images which is used in the laboratory is based on the severity scales suggested by the international clinical standards and diabetic macular edema disease scales [25]. The neural networks (NNs) are analytical paradigms that have variables used in particular estimations, like convolutions and totalization. In general, the NN is gauged with multiple layers having an input and hidden layers from which it is used to estimate an output, for instance, a class of DR as shown in Figure 2.6. The variables used in calculating the end-result can be altered on task-based mode by reducing the fault between NN outputs and the physical observations.

Earlier diagnosis of this eye disease can be managed and treated using fundus or retinal screening tests. A manual process to detect this eye disease is painful due to the testing tools deficiency and proficiency. The disease is absent to show the symptoms at the initial stage because of that the eye-specialists require a considerable time to explore the fundus images that causes a delay in medications.

Normal eye   Background retinopathy   Pre-proliferative Retinopathy   Proliferative Retinopathy   Maculopathy   Evidence of multiple Laser scars

FIGURE 2.6   Eye images of diabetic retinopathy based on grading [24].

2.5.2 ELECTRONIC HEALTH RECORDS (EHRS)

EHR [26] is the digitalization of patients' health information which includes pre-and post-historic conditions of the patient having an illness, their treatment, medications, and doctor's prescriptions. Whenever the patient visits different physicians, their EHR must be updated to get a long-term view of the patients' diseases and health conditions. Hence, mining into the EHRs can boost the decision-making procedure and the communication of assessments via electronic resources to others who are all involved. EHR helps the physician for creating, maintaining, revealing, and manipulating the database of the private files. These datasets are stored in a clinical repository, one of the wide-ranging datasets of the medical field that will assist clinical and medical research. This could also support to keep large-scale monitoring for contaminations and generate alert messages at the beginning of the diseases once the symptoms appear. On the other hand, collecting details about private data also constitutes the possibility of intrusion of one's privacy through social media platforms like WhatsApp, Facebook, Twitter, or Instagram. According to the current situation, it gives a warning for individuals who can govern their private health information while generating mechanisms to attain more information for a wide range of studies using DL algorithms.

2.5.3 GENOMICS

From the time when DL was introduced, it has influenced as one of the favorable and exciting fields of genomics where they are consistently used to report a range of questions stretching from the indulgent of protein binding from DNA sequences, epigenetic alterations [27–30], forecasting gene-expression from epigenetic characters [31], or forecasting the separation of individual cells [32]. From the DL, perspective genomics is a sub-domain in which it has altered the traditional Conventional Neural Network (CNN) integrated with unique (non-image or non-temporal) data illustrations. Contemporary genomic technologies gather a wide range of measurements ranging from a person's DNA sequence to the number of numerous proteins present in their plasma. The distinctive pipeline for constructing a deep-learning framework in genomics includes: creating raw data (gene data), transforming this raw data into input data by using data tensors, and feeding these tensors through NNs which then governs particular biomedical applications (Figure 2.7). When the clinicians have

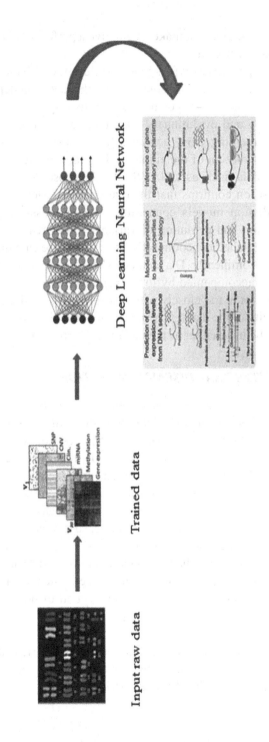

FIGURE 2.7 Deep learning in genomics.

ample knowledge of genomes, it makes them give a perfect diagnosis and, in turn, results in better treatment.

A crucial task for physicians is to define any relevancy with the patient's genome to diagnose the disease. To some extent, this decision depends on predicting the pathogenicity – a potential ability that uses structures like protein configuration and evolutionary maintenance to training learning algorithms. Data-streaming from genomes too behave as a biomarker for the arrival and evolution of the ailments. The ease of use of veritable treasure troves of data such as DNA, RNA, chromosome interactions, methylation, and so on confirms that there are adequate learning sets to form precise prediction models concerning gene expression, genomic sequence, or collation and evolution of variants. Other features such as documentation of stretched noncoding RNAs or splice-site calculation can also be examined. Prominently, DL methods should be matched with the current ML models accompanied by lower parameters to certify that the intricacy of DL has not fit too closely with inaccurate data. Independent of the class and capacity of the datasets being examined and the doubts being requested, DL can be used either as a recipient or as an ambiguity.

2.5.4   GASTROINTESTINAL DISEASES DETECTION

In the current era, DL has attained notable attention amongst the existing medical image analysis methods which provide analogous results as humans in classification-related problems. The gastrointestinal zone comprises all organs starting from mouth to anus, that intricate dissolve the food and absorb the nutrients. The organs encompassed digestive tracts are the esophagus, stomach, duodenum, large bowel, and small bowel. Esophagus, stomach, and duodenum organized the top gastric area while large and small intestines organized the bottom gastric area.

The consumption and concentration of foods get affected by the ailments like infection, hemorrhage, contaminations, and lesions in the gastric area. Sore is the main ground of hemorrhage in the top gastric area and cyst, tumor, or inflammation are the main ground of hemorrhage in the large bowel and hemorrhage due to atypical veins are the barriers for small bowel.

The most popular radioscopic equipment, including gastric endoscopy, wireless capsule endoscopy, colonoscopy, and tomography, play a vibrant role in detecting the illness related to the gastric area.

By putting altogether from Table 2.2, a drive to excel DL implementations were observed which is linked with endoscopic, pathological, and CT colonography [43–45] images pooled with transferal knowledge is compared with historical classification or detection algorithms.

TABLE 2.2   Literature Survey of the Gastrointestinal Disease Detection

Authors	Dataset Used	Methodology	Objective
Lee et al. [33]	Clinical data	Survival recurrent network	Predict survival rate after surgery
Lee et al. [34]	Gastric endoscopic image	ImageNet	Detect malignancies
Nakashima et al. [35]	Endoscopic images from ImageNet large scale visual recognition challenge (ILSVRC) 2004	GoogleNet fine-tuning	*Helicobactor pylori (Hpylori)* infection
Li et al. [36]	Gastric endoscopic images	GastricNet	Gastric cancer detection
Li et al. [37]	BOT gastric slice dataset	GT-Net	Automatic gastric tumor segmentation
Hirasawa et al. [38]	Endoscopic images of gastric cancer	Convolution neural network (CNN)	Malignancy detection
Xiao and Max [39]	Wireless capsule endoscopy (WCE) images	Deep convolutional neural network (DCNN)	Gastrointestinal bleeding detection
Panpeng et al. [40]	Wireless capsule endoscopy (WCE) images	Deep convolutional neural network (DCNN)	Intestinal hemorrhage
Diana et al. [41]	Small bowel capsule endoscopy (SBCE) images of small intestine	Deep convolutional neural network (DCNN)	Detect gastrointestinal angiectasia (GIA)
Gregor et al. [42]	Colonoscopy images	Deep convolutional neural network (DCNN)	Localize and detect polyps in colonoscopies

2.5.5   CARDIAC IMAGING

As reported by the WHO, cardiovascular diseases (CVDs) are the deadliest disease universally and the total number of individuals affected due to it increases exponentially. MRI and CT scans are the highest acceptable mechanism for heart images in the recent MI modalities. The most arduous task done manually in cardiac CT scans is the one that measures coronary artery calcium (CAC) scoring – is a test to calculate the amount of calcium deposited in the heart walls. By measuring it manually, makes it to be an unbearable task for epidemiological studies as in Ref. [46].

From Figure 2.8 it depicts that the inputs fed up to the DL algorithms may vary from the heart CT images, heart rate, and whatever information relevant to the heart disease and the output obtained from the DL algorithm is how much percentage the individual having a risk assessment for CVD.

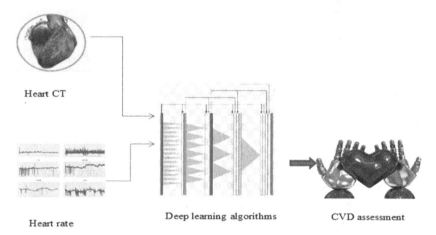

FIGURE 2.8   Deep learning frameworks for cardiovascular disease.

2.5.6   ALZHEIMER'S AND PARKINSON'S DETECTION

Neuroimaging play a major role in understanding brain-related syndrome for the past decades. MRI scan is one of the imaging modalities that must have cutting-edge performance over the other imaging modalities. In whatever way the attained neuroimaging will be, some similarities in disease prototype make it very challenging to discriminate such disorders precisely. AD is a step-by-step neurological chaos in cerebrum cells, and the impact is a continuous decline in thinking skills obstructing the ability to convey day-to-day activities. The correct detection of AD in the earlier phase plays a significant role in patients' care. Parkinson's disease is a neurological chaos produced due to the scarcity of dopamine generating neurons, which blocks the substantia nigra region responsible for the motion of the body. The common symptoms are shivering in the hands, legs, or head, followed by the slowness of movement, stiffness, and impaired balance. A patient whoever having these neurological disorders will be a burden to themselves and their families. For that cause, it is predominant to treat these disorders in the earlier phase even before their progression can be slowed down [47].

2.6 A NOVEL PROTOTYPE FOR COMPUTER-AIDED DIAGNOSTIC (CAD) SYSTEM

CAD has evolved into an emergent technology in the healthcare field until now from the earlier 20th century. Physicians frequently go for diagnostic tools such as an X-ray, CT scan, or MRI scan for diagnosing anomalies or illness. Even though health practitioners have a piece of medical knowledge and decision to confirm and conclude the diseases, therapeutic imaging is a factor to authenticate any diagnosis. It cannot replace the medical practitioner but it can help to decide further treatment and futuristic plans. CT and MRI scans permit the doctor to observe the efficacy of medical care and fine-tune the procedures whenever needed. In Figure 2.9,

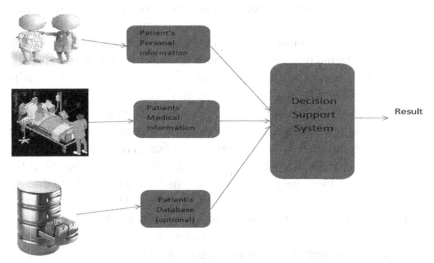

FIGURE 2.9 A novel prototype for a CAD system.

it was clearly shown that for every decision support / CAD system, both the initial and ancillary information are necessary and collected from the victim. This information is stored in a database for future requirements. Based on the computerized and non-computerized data gained from the victim and the set of rules they have used; the CAD system's accuracy will differ. A potential approach has to be inclined to design the CAD system to raise the efficacy and dropdown the processing time by assisting the physician to detect too small anomalies – those who are not seen through our naked eyes.

2.7 ETHICAL IMPLICATIONS AND FUTURE INFERENCES OF DL-BASED-CAD

During this 21st century, data construction is extensive and the curve for the usages of this data has grown steadily. One of the major areas producing a huge amount of data in healthcare. The current DL application helps healthcare professionals improve decision-making, pinpoint anomalies, and improve the efficacy of clinical trials and research. One of the keys confronts while learning MI is eradicating the noise in digital images [13]. It is essential that medical images be should be clear, flawless, and noise-less. Even though noise gives an undesirable look to an image, the noteworthy challenge of noise is that it can cover and diminish firm features' brightness within the image. In researchers' opinion, while reducing image noise the "screen" for denoising is raised to some degree, and precisely the low contrast objects within the image become visible.

Undoubtedly the most problematic issue of today's technologies is transparency. In today's technology, once the patients' information is digitized, the hackers/unknown persons can know about the patient's details without their knowledge. To overcome this, network security must be included in the CAD system and their personalized gadgets.

Another important issue for the existing DL algorithms is that they are ineffective to interpret or to explain the medical terms [48]. In general, neither DL algorithms nor the CAD systems which are responsible for the diagnosis cannot give proper explanations for the cause of the disease to the patients or their relatives. For instance, if a CAD system diagnoses the patient having cancer based on the MI records, he/she will expect the reason to know why. Ultimately, the causal nature of human beings cannot be replaced by a machine still it was super intelligent.

A DL-based system cannot operate in 'one-algorithm-fit-all,' as each application needs different algorithms to execute it [49]. In the future, DL could proactively propose algorithms that could modify automatically based on the requirements.

The forthcoming healthcare field has never reached such a mount peak through AI and ML, which allows developing solutions that outfit for particular needs within the healthcare industry, but DL in healthcare can become amazingly powerful by giving a second opinion to the clinicians and transforming patient care. Ultimately, DL-based systems support the medical professionals and streamline and simplify complex data analysis by learning them and improving diagnosis. As a result of this, it can reduce reporting delays and progress workflows.

The next big question we've to point out is its achievement over a human being, but it is not easy. Building a cleverness that surpasses human levels is outlying from the real situation, as we believe. The world is complex and open-ended, and the highly specialized narrow technologies we have today can't be trusted entirely yet for those tasks where this complexity is fully displayed, as in personalized medicine.

KEYWORDS

- artificial intelligence
- computed tomography
- computer-aided diagnosis
- deep learning
- magnetic resonance imaging
- positron emission tomography
- three-dimensional

REFERENCES

1. Doi, K., (2007). Computer-aided diagnosis in medical imaging: Historical review, current status and future potential. *Computerized Medical Imaging and Graphics: The Official Journal of the Computerized Medical Imaging Society, 31*(4, 5), 198–211.
2. European Society of Radiology, (2019). What the radiologist should know about artificial intelligence: An ESR white paper. *Insights into Imaging, 10*(1), 44.

3. Takahashi, R., & Kajikawa, Y., (2017). Computer-aided diagnosis: A survey with bibliometric analysis. *International Journal of Medical Informatics, 101*, 58–67.
4. Lodwick, G. S., Keats, T. E., & Dorst, J. P., (1963). The coding of roentgen images for computer analysis as applied to lung cancer. *Radiology, 81*, 185–200.
5. Becker, H. C., Nettleton, W. J. Jr., Meyers, P. H., Sweeney, J. W., & Nice, Jr. C. M., (1964). Digital computer determination of a medical diagnostic index directly from chest X-ray images. *IEEE Trans. Biomed Eng., 11*, 67–72.
6. Meyers, P. H., Nice, C. M. Jr., Becker, H. C., Nettleton, Jr. W. J., Sweeney, J. W., & Meckstroth, G. R., (1964). Automated computer analysis of radiographic images. *Radiology, 83*, 1029–1034.
7. Winsberg, F., Elkin, M., Macy, J. Jr., Victoria, B., & Weymouth, W., (1967). Detection of radiographic abnormalities in mammograms through optical scanning and computer analysis. *Radiology, 89*, 211–215.
8. Toriwaki, J., Suenaga, Y., Negoro, T., & Fukumura, T., (1973). Pattern recognition of chest X-ray images. *Comput. Graph Image Process, 2*, 252–271.
9. Roellinger, F. X. Jr., Kahveci, A. E., Chang, J. K., Harlow, C. A., Dwyer, S. J. III., & Lodwick, G. S., (1973). Computer analysis of chest radiographs. *Comput. Graph Image Process, 2*, 232–251.
10. Chan, H. P., Doi, K., Galhotra, S., Vyborny, C. J., MacMohan, H., & Jokich, P. M., (1987). Image feature analysis and computer-aided diagnosis in digital radiography. I. Automated detection of microcalcifications in mammography. *Med. Phy., 14*, 538–548.
11. Fujita, H., Doi, K., Fencil, L. E., & Chua, K. G., (1987). Image feature analysis and computer-aided diagnosis in digital radiography. *Med. Phy., 14*, 549–556.
12. Katsuragawa, S., Doi, K., & MacMahon, H., (1988). Image feature analysis and computer-aided diagnosis in digital radiography: Detection and characterization of interstitial lung disease in digital chest radiographs. *Med. Phy., 15*, 311–319.
13. Yang, W., & Haomin, Z., (2006). Total variation wavelet-based medical image denoising. *International Journal of Biomedical Imaging, 2006*.
14. Koitka, S., Demircioglu, A., Kim, M. S., Friedrich, C. M., & Nensa, F., (2018). Ossification area localization in pediatric hand radiographs using deep neural networks for object detection. *PLoS One, 13*(11), e0207496.
15. Deniz, C. M., Xiang, S., Hallyburton, R. S., Welbeck, A., Babb, J. S., Honig, S., Cho, K., & Chang, G., (2018). Segmentation of the proximal femur from MR images using deep convolutional neural networks. *Sci. Rep., 8*, 16485.
16. Abd-Ellah, M. K., Awad, A. I., Khalaf, A. A. M., & Hamed, H. F. A., (2018). Two-phase multi-model automatic brain tumor diagnosis system from magnetic resonance images using convolutional neural networks. *EURASIP J. Image Video Process, 97*, 1–10.
17. Konstantinos, K., Christian, L., Virginia, F. J. N., Joanna, P. S., Andrew, D. K., David, K. M., Daniel, R., & Ben, G., (2016). Efficient multi-scale 3D CNN with fully connected CRF for accurate brain lesion segmentation. *Medical Image Analysis, 36*, 61.
18. (2021). *Latest Trends in Medical Monitoring Devices and Wearable Health Technology.* https://www.businessinsider.in/science/latest-trends-in-medical-monitoring-devices-and-wearable-health-technology/articleshow/70295772.cms (accessed on 14 May 2022).

19. (2021). *Deep Learning in Healthcare: How it's Changing the GAME*. https://www.aidoc.com/blog/deep-learning-in-healthcare/ (accessed on 14 May 2022).

20. Suzuki, K., Shiraishi, J., Abe, H., et al., (2005). False-positive reduction in computer-aided diagnostic scheme for detecting nodules in chest radiographs by means of massive training artificial neural network. *Acad. Radiol., 12*, 191–201.

21. Van, G. B., Ter Haar, R. B. M., & Viergever, M. A., (2001). Computer-aided diagnosis in chest radiography: A survey. *IEEE Trans. Med. Imaging, 20*, 1228–1241.

22. Chen, S., Suzuki, K., & Mac, M. H., (2011). Development and evaluation of a computer-aided diagnostic scheme for lung nodule detection in chest radiographs by means of two-stage nodule enhancement with support vector classification. *Med. Phys., 38*, 1844–1858.

23. Suzuki, K., & Armato, S. G. 3rd., Li, F., et al., (2003). Massive training artificial neural network (MTANN) for reduction of false positives in computerized detection of lung nodules in low-dose computed tomography. *Med. Phys., 30*, 1602–1617.

24. Zachariah, S., Wykes, W., & Yorston, D. (2015). Grading diabetic retinopathy (DR) using the Scottish grading protocol. Community Eye Health, 28(92), 72, 73.

25. Wilkinson, C. P., et al., (2003). Proposed international clinical diabetic retinopathy and diabetic macular edema disease severity scales. *Ophthalmology, 110*(9), 1677–1682.

26. Paolo, C., Emanuela, L., Claudio, M., Lucia, F., Agostino, M., Walter, R., & Maria, L. S., (2016). The impact of electronic health records on healthcare quality: A systematic review and meta-analysis. *European Journal of Public Health, 26*(1), 60–64.

27. Alipanahi, B., Delong, A., Weirauch, M. T., & Frey, B. J., (2015). Predicting the sequence specificities of DNA-and RNA-binding proteins by deep learning. *Nat. Biotechnol., 33*, 831.

28. Zhou, J., & Troyanskaya, O. G., (2015). Predicting effects of noncoding variants with deep learning-based sequence model. *Nat. Methods, 12*, 931.

29. Kelley, D. R., Snoek, J., & Rinn, J. L., (2016). Basset: Learning the regulatory code of the accessible genome with deep convolutional neural networks. *Genome Res., 26*, 990–999.

30. Kelley, D. R., Yakir, A. R., Bileschi, M., Belanger, D., McLean, C. Y., & Snoek, J., (2018). Sequential regulatory activity prediction across chromosomes with convolutional neural networks. *Genome Res., 28*, 739–750.

31. Singh, R., Lanchantin, J., Robins, G., & Qi, Y., (2016). Deep chrome: Deep-learning for predicting gene expression from histone modifications. *Bioinformatics, 32*, 639–648.

32. Angermueller, C., Lee, H. J., Reik, W., & Stegle, O., (2017). DeepCpG: Accurate prediction of single-cell DNA methylation states using deep learning. *Genome Biol., 18*, 67.

33. Lee, J., An, J. Y., Choi, M. G., et al., (2018). Deep learning-based survival analysis identified associations between molecular subtype and optimal adjuvant treatment of patients with gastric cancer. *JCO Clinical Cancer Informatics, 2*, 1–14.

34. Lee, J. H., Kim, Y. J., Kim, Y. W., et al., (2019). Spotting malignancies from gastric endoscopic images using deep learning. *Surg. Endosc., 33*, 3790–3797.

35. Nakashima, H., Kawahira, H., Kawachi, H., et al., (2018). Artificial intelligence diagnosis of Helicobacter pylori infection using blue laser imaging-bright and linked color imaging: A single-center prospective study. *Ann. Gastroenterol., 31*, 462.

36. Li, Y., Li, X., Xie, X., et al., (2018). Deep learning-based gastric cancer identification. In: *2018 IEEE 15ʰ International Symposium on Biomedical Imaging (ISBI 2018)* (pp. 182–185).

37. Li, Y., Xie, X., Liu, S., et al., (2018). Gt-net: A deep learning network for gastric tumor diagnosis. In: *2018 IEEE 30ʰ International Conference on Tools with Artificial Intelligence (ICTAI)* (pp. 20–24).

38. Hirasawa, T., Aoyama, K., Tanimoto, T., et al., (2018). Application of artificial intelligence using a convolutional neural network for detecting gastric cancer in endoscopic images. *Gastric Cancer, 21,* 653–660.

39. Xiao, J., & Max, Q. H M., (2016). A deep convolutional neural network for bleeding detection in wireless capsule endoscopy images. *Annual International Conference of the IEEE Engineering in Medicine and Biology Society, 639*–642.

40. Panpeng, L., Ziyun, L., Fei, G., Li, W., & Jun, Y., (2017). Convolutional neural network for intestinal hemorrhage detection in wireless capsule endoscopy images. In: *2017 International Conference on Multimedia and Expo* (pp. 1518–1523).

41. Diana, E. Y., John, N. P., Romain, L., Xavier, D., & Anastasios, K., (2020). Poor quality of small bowel endoscopy images has a significant negative effect in the diagnosis of small bowel malignancy. *Clinical and Experimental Gastroenterology, 13,* 475–484.

42. Gregor, U., Priyam, T., Talal, A., Mohit, M., Farid, J., William, K., & Pierre, B., (2018). Deep learning localizes and identifies polyps in real-time with 96% accuracy in screening colonoscopy. *Gastroenterology, 155*(4), 1069–1078.

43. Suzuki, K., Hori, M., McFarland, E., et al., (2009). Can CAD help improve the performance of radiologists in the detection of difficult polyps in CT colonography? *Proceedings of RSNA Annual Meeting.* Chicago, IL.

44. Dachman, A. H., Obuchowski, N. A., Hoffmeister, J. W., et al., (2010). Effect of computer-aided detection for CT colonography in a multi-reader, multicase trial. *Radiology, 256,* 827–835.

45. Petrick, N., Haider, M., Summers, R. M., et al., (2008). CT colonography with computer-aided detection as a second reader: Observer performance study. *Radiology, 246,* 148–156.

46. Geert, L., Francesco, C., Jelmer, M W., Bob, D. D. V., Tim, L., Jonas, T., & Ivana, I., (2019). State-of-the-art deep learning in cardiovascular image analysis. *Journal of American College of Cardiology (JACC) Cardiovascular Imaging, 12*(8), 1549–1565.

47. Manan, B. T. N., Nasrat, Z. Z., Shamim, K. M., Shamim Al, M., & Mufti, M., (2020). Application of deep learning in detecting neurological disorders from magnetic resonance images: A survey on the detection of Alzheimer's disease, Parkinson's disease and schizophrenia. *Brain Informatics, 7,* 11.

48. *Deep Learning & Healthcare: All That Glitters Aren't Gold.* (2021). https://towardsdatascience.com/deep-learning-in-healthcare-all-the-glitters-aint-gold-4913eec32687 (accessed on 14 May 2022).

49. *Improving Diagnosis in Healthcare.* (2015). https://nap.nationalacademies.org/download/21794 (accessed on 1 June 2022).

CHAPTER 3

Extensive Study of WBC Segmentation Using Traditional and Deep Learning Methods

CHANDRADEEP BHATT,[1] INDRAJEET KUMAR,[1]
SANDEEP CHAND KUMAIN,[1] and JITENDRA KUMAR GUPTA[2]

*[1]CSED, Graphic Era Hill University, Dehradun, Uttarakhand, India,
E-mails: bhattchandradeep@gmail.com (C. Bhatt),
erindrajeet@gmail.com (I. Kumar), skumain@gehu.ac.in (S. C. Kumain)*

*[2]CSED, GRD Institute of Management and Technology, Dehradun,
Uttarakhand, India, jk760429@gmail.com*

ABSTRACT

According to the WHO annual report, the death rate related to blood diseases is very high in the Asian continent. The existing traditional system is prolonged and tedious and also based on the expertise's knowledge. Therefore, the development of an automated blood-related disorder diagnostic system is very essential to make the system error-free and more effective. As per the hematologist's opinion, most of the disease can be identified by the White blood cells (WBC) related information. Thus, the main goal of this chapter is to segment WBCs from microscopic images using different traditional and deep learning (DL) algorithms. For this work, three traditional methods, i.e., global thresholding, k-means clustering and one DL-based model, i.e., U-Net is implemented. The extensive work of experimentations has been conducted on ALL-IDB dataset. The

Application of Deep Learning Methods in Healthcare and Medical Science.
Rohit Tanwar, PhD, Prashant Kumar, PhD, Malay Kumar, PhD, & Neha Nandal, PhD (Editors)
© 2023 Apple Academic Press, Inc. Co-published with CRC Press (Taylor & Francis)

performance of the Global thresholding, the K-means clustering technique and the U-Net method is evaluated with the help of metrics like Jaccard Index, accuracy, sensitivity, and specificity. It has observed that the U-Net method is the best performing method for the WBC segmentation.

3.1   INTRODUCTION

Generally, all blood-related diseases such as infection, inflammation, allergies, and leukemia initially identified by examining the blood smear. Blood is made up of Red-Blood-Cells (RBCs), White-Blood-Cells (WBCs), Plasma, and Platelets. WBCs account for only about 1% of human blood but its impact is big. WBCs are element of the immune system, and these blood cells protect the body against viruses, bacteria, and other diseases [1]. The bone marrow's stem cells are the main source of WBCs and more than 80% WBCs are present in bone marrow. The major components of microscopic blood smear image are shown in Figure 3.1.

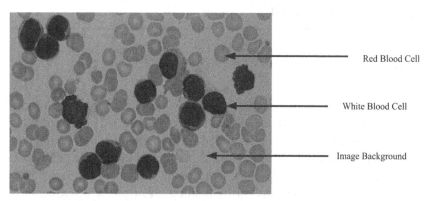

FIGURE 3.1   Components of microscopic blood smear image.

White cells are broadly classified into two categories based on the presence of particle as granulocytes and agranulocytes [2]. Agranulocytes are WBCs that have very small granules which contain proteins and others are granulocytes. Basically, granulocyte cells are categorized into three types. (*i*) Basophils: These cells makes up account nearly 1% of total WBCs available in the human body and that are mostly found in greater than before in case of total count after met an allergic reaction. (*ii*) Eosinophils:

It is answerable for vermin reason. These cells are also responsible for immune responses and inflammatory reaction in the entire body as well. (*iii*) Neutrophils: Most of the WBCs are neutrophils and the cells kill and digest fungi and bacteria.

Basically, we have two types of agranulocytes cells. (i) Lymphocytes: These WBCs produce antibodies to fight against viruses and other harmful bacteria. These cells have three sub-types as: T-cells, B-Cells, and natural killer cells. T-type cells are helpful to identify and removing infection origin cells and B-type cells generate antibodies to boost the immunity against infections. Attacking and destroying the viral cells or cancerous cells are done by natural killer cells. (ii) Monocytes: Occurrence of these cells lies between 2% and 8% of total WBCs count in the body. Monocytes cells have a longer lifespan than other WBCs and these are normally found when the body fights against chronic infection. Sample images of WBC classifications are shown in Figure 3.2.

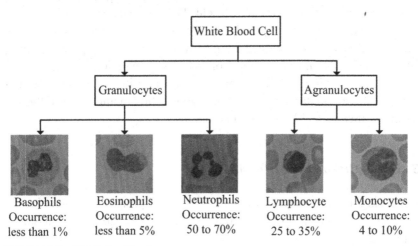

FIGURE 3.2 WBC classification.

3.1.1 PROBLEMS RELATED TO WBCS

The normal range of WBCs measure in per cubic millimeter and it is generally based on age. The normal range for adult should be 4,500 to 11,000 and for newborn infant is 13,000 to 38,000. Besides this, WBC count can be high or low due to number of reasons. If the human's body is producing more WBCs than the normal range, the person may suffer from diseases like asthma, heart attack, leukemia, etc. It is possible that a person's body can produce fewer WBCs than the normal range; the person can suffer from bone marrow disorder, leukemia, lymphoma, etc. These diseases can be identified by performing a WBC count using a blood test. During a manual examination, a lab technician performs a blood test and examines the count of WBCs in the blood is generally a part of the Complete Blood Cell (CBC) test. This test is useful to look body disorders such as infection, inflammation, allergies, and leukemia. For diagnose these diseases and other blood-related diseases, identification of WBCs is the very first step that are usually done by hematologist using an optical microscope.

3.2 DATA METHODOLOGY AND DATASET

Since physical examination of WBC is time consuming, complex, expensive as well as needs experts in this field. Thus, automated computer-aided diagnosis (CAD) system that helps hematologist in the diagnostic process can be so useful [3, 4]. Segmentation of WBC is a first step for developing an automated diagnosis system. For WBC segmentation, we have a lot of machine learning and deep learning (DL) algorithms [5]. This section presents three popular techniques for segmentation, namely global thresholding, k-means, and U-net DL method.

3.2.1 GLOBAL THRESHOLDING

According to global thresholding algorithm [6, 7] for image, each pixel of the image is mapped into one group among the two groups discussed in Eqn. (1):

$$G_1 = P, \ if \ 0 \le P < th$$
$$G_2 = P, \ if \ th \le P < M - 1 \tag{1}$$

where; P represent single-pixel of mage in M gray-level, M = {0, 1, 2, …, 255}, threshold value is denoted by *th*, while the two pixel groups are denoted by G1 and G2. The above method can be expanded by introducing two threshold values {th_1, th_2} and every pixel of the image is mapped into one group among the three groups as denoted in Eqn. (2):

$$G_1 = P, \ if \ 0 \le P < th_1$$
$$G_2 = P, \ if \ th_1 \le P < th_2 \qquad (2)$$
$$G_3 = P, \ if \ th_2 \le P < M - 1$$

The steps involving in global thresholding method are illustrated in following four steps:

- **First Step:** In this step, LAB color space image $I_g(x, y)$ is created from RGB color-space E(x, y) image.
- **Second Step:** The histogram H image is constructed from gray-level image $I_g(x, y)$.
- **Third Step:** By using global thresholding method [8], calculate the two threshold cost th_1 and th_2 from histogram H image so that it fulfills the condition given in Eqn. (3):

$$th_1, th_2 = Arg\left(Max_{\min(H) \le th \le \max(H)} \left(\sigma^2 (th) \right) \right) \qquad (3)$$

In Eqn. (3), the threshold value from histogram H is denoted by th_1 and th_2, while σ is inter-class variance. The minimum and maximum gray-level intensity is denoted by min(H) and max(H) in histogram H.

- **Fourth Step:** Apply thresholding by putting the value of th_1 and th_2 into Eqn. (2).

3.2.2 K-MEAN ALGORITHM

The K-means technique is a popular clustering algorithm that totally depends on the Euclidean distance. It uses the Euclidean distance as the prime constraint of homogeneity, that refers to the smaller space is the more homogenous for the both objects. The first step of this clustering method is to divide all elements into k number of groups. Every element normally fit into the group having the closest mean value, which provides the sample to the group. Author Zhang et al. [9] is used this characteristic to divide $N_R \times N_C$ pixels of the object into various groups according to their intensity level. The N_R is used to represent the number of rows in the object, and N_C is used to denote the number of columns in the object. A single pixel in the object represented as $\{x^{(1)}, x^{(2)}, \dots x^{(P)}$, where $x^{(i)} \varepsilon R^n$, i ε [1 P], $P = N_R \times N_C$.

The main two steps of k-means clustering technique are illustrated as:

*   Choose the first group of k number of cluster centroids c_1, c_2, ..., $c_k \, \varepsilon \, R^n$ arbitrarily.
*   Once the first k-cluster centroids are selected, the method goes on by blinking between first two ladders till the convergence.

1. **Assignment Step:** The objective of this step is to find the cluster for each sample $x^{(i)}$ by computing the Euclidean distance of sample $x^{(i)}$ from each group's centroid c_j ($j \varepsilon [1 \; k]$) as:

$$C^{(i)} = \arg m \, in \, \| x^{(i)} - c_j \|^2 \qquad (4)$$

where; $C^{(i)}$ denotes the cluster number of sample $x^{(i)}$.

2. **Update Step:** Recomputing the new mean value of the group centroid of the elements in the new groups is done in this step. Recomputing function is given in Eqn. (5).

$$c_j = \frac{\sum_{i=1}^{P} 1\left\{c^{(i)} = j\right\} x^{(i)}}{\sum_{i=1}^{P} 1\left\{c^{(i)} = j\right\}} \qquad (5)$$

The k-means and Hidden Markov Random Field (HMRF) is applied to developed 2-stage of segmentation process [10]. These techniques are used to group several types of Acute Myeloid Leukemia (AML) cells from bone marrow images. This segmentation technique achieved average accuracy greater than 96%. A soft covering-based rough k-means method [11] can be used to segment the leukemia nucleus images. This algorithm identifies the highest value from the histogram image and utilizes it for defining the number of cluster k.

3.2.3   U-NET DEEP LEARNING MODEL

The U-Net is a Convolutional Neural Network-based network which is widely used for doing rapid and accurate segmentation of biomedical image [12]. The architecture of U-Net involving three phases: the down-sampling phase or decoding phase, bottleneck, and the up-sampling phase or encoding phase.

The down-sampling phase is made of several contraction block, and each block contains two 3 × 3 convolution layer (CL) which is followed by the activation function 'Rectified Linear Unit' (ReLU) and the max pooling operation for down-sampling. After each block the number feature get doubles so that the network can understand the complex structures accurately. At every pooling action, there must be losses of some important spatial information and the losses can be managed by using skipping connections [13]. The architecture of U-Net DL network is shown in Figure 3.3.

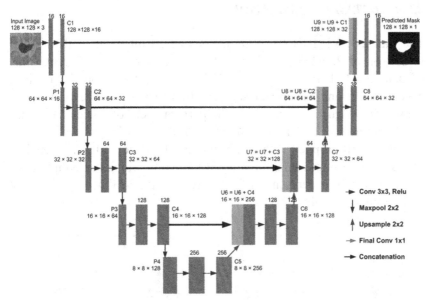

FIGURE 3.3   U-net architecture.

The bottleneck layer works as a connector between decoding and encoding network. This connector uses two 3 ×3 CL stepped up by 2 × 2 up CL. Identical to down-sampling phase, up-sampling phase consist lot of expansion blocks as well. The all block of expansion contain two 3 × 3 CLs trailed by a 2 × 2 up CLs. Later each up-sampling layer, the numbers of features gets half to maintain symmetry. The number of contraction block involved in down-sampling phase is equal to a number of expansion blocks in up-sampling phase. After that, final mapping follow-out another

3 × 3 CLs with the required amount of feature vectors to the required amount of desired segments.

The U-Net model makes use of ReLU activation function at each CLs excluding the final layer [14]. At final layer, the sigmoid activation function is used to handle the binary problem. Actually, the entire model is used for two class classification problem, so the binary cross-entropy loss function is suitable. Finally, the learning rate and network layer weight is managed by *Adam* optimizer.

3.2.4   DATASET PREPARATION

The dataset ALL-IDB has ROI images of 24-bit depth with original resolution of 2,592×1,944. This database contains two different kinds of dataset (ALL-IDB1 and ALL-IDB2) concentrated on the nucleus classification and the nuclei segmentation task [15]. The ALL-IDB2 dataset is designed for analyzing the performances of nuclei segmentation techniques. This dataset is a group of images belongs to ALL-IDB1 database with cropped area of normal and blast cells. It has total 260 blood image samples and among them mostly belongs to lymphoblast class. Figure 3.4 represents a few images of the healthy and the blast cells from ALL-IDB2 dataset.

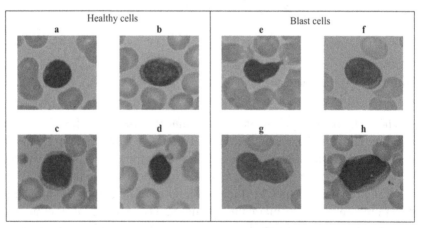

FIGURE 3.4   Images taken from ALL-IDB2 – (a–d) healthy class images, (e–h) blast cell images.

3.3 PROPOSED WORK

The proposed work contains general steps for nucleus segmentation given in Figure 3.5. It consists of image resizing phase, normalization process, general segmentation techniques and post-processing phase.

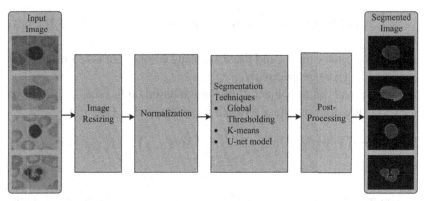

FIGURE 3.5   Workflow diagram.

The proposed model operates on fixed-size of input images. Therefore, every input images has to be resized using image resizing operation. The image resizing phase includes image resizing operation on all the input images. In this database, all images are different size. So initially, it has been converted to fix size of images. All Input images are converted to 128 × 128 × 3 dimensions. The proposed model also required mask for every sample, therefore the mask is also generated. The size of mask images is also 128 × 128 × 1. The next step of preprocessing step is normalization. Normalization gives the assurance that each input parameter data is equally distributed in the range of intensity pixel values. The normalization is performed by dividing each input pixels value by 255 to scale the input intensities value in the range of 0 to 1. By using this, the saturation of the sigmoid function before time is prevented. After normalization, the entire dataset is bifurcated into the training and testing instance of image set, and then the training dataset is passed to the segmentation module. The segmented mask is generated by the proposed model.

After this, the generated mask with respect to the input image is processed by post-processing operation. In post-processing operation, element wise matrix multiplication is performed and the nucleus part of the input images is finally obtained as resultant image. Before this multiplication, the morphological operation [16] is performed on the generated mask for the removal of noise and unwanted portions. Morphological opening (MO) based on erosion is used and the erosion function E(x,y) is defined in Eqn. (6) for an input image i(x,y).

$$E(x, y) = i(x, y) \ominus SC(x, y) = \left\{ a : SC(x, y)_a \subseteq \left\{ i(x, y) \right\} \right. \quad (6)$$

Here the binary image and the structuring component is represented by i(x,y) and SC(x,y), respectively.

Then spread out the object of the image based on the size and shape of SC(x,y) using dilation D(x,y), which is represented in Eqn. (7).

$$D(x, y) = i(x, y) \oplus SC(x, y) = \left\{ a : \left[\left(SC(x, y)_a \cap i(x, y) \right] \subseteq i(x, y) \right. \right.$$
$$(7)$$

The morphological opening (MO) function can be represented mathematically as given in Eqn. (8).

$$MO = E(x, y) \circ D(x, y) = \left\{ i(x, y) \ominus E(x, y) \right\} \oplus D(x, y) \quad (8)$$

3.4   EXPERIMENTAL DESCRIPTION AND RESULT ANALYSIS

3.4.1   EXPERIMENTAL SETUP

Overall experimentation has been performed at a workstation. The system configuration of the workstation is having Intel Xeon W-2014 CPU @ 3.2 GHz, 64-GB RAM, 4-GB NVIDIA graphics, 256 GB SSD and 2TB SATA HDD. The entire dataset is kept on the same workstation and all experiments have been performed under Python environment.

3.4.2   DESCRIPTION OF EXPERIMENTS

Description of all three experiments is shown in Table 3.1 with their purpose and number of samples.

TABLE 3.1 Description of Experiments

Experiment No.	Description
Experiment 1	WBC nucleus segmentation using global thresholding method
Experiment 2	WBC nucleus segmentation using K-means
Experiment 3	WBC nucleus segmentation using the U-net model

To compute the effectiveness of proposed U-Net nucleus segmentation model, a dataset of random 20 images from ALL-IDB2 have been taken. The performance is evaluated on the basis of values of quantitative parameters such as sensitivity, Jaccard index, and accuracy [16]. The sensitivity parameter is the percentage of pixels test genuinely positive to that give positive result and therefore, it is also called true positive rate. Specificity means the percentage of pixels test genuinely negative to that give negative result and is also called true negative rate. The output of the proposed system is given in Figure 3.6.

The following Eqns. (9) and (10) is used to compute the sensitivity and the specificity, respectively.

$$Sensitivity(\%) = \left[\frac{TP}{TP+FN}\right] \times 100 \tag{9}$$

TP means the number of pixels appearing in golden truth segmented region as well as the segmented region marked by the model. FN means the number of pixels appearing in the golden truth segmented region but not present in the segmented region detected using the developed model.

$$Specificity(\%) = \left[\frac{TN}{TN+FP}\right] \times 100 \tag{10}$$

TN means the number of pixels absent in ground truth segmented region as well as in segmented region detected using the developed model. In Eqn. (10), the FP represents the number of pixels absent in golden truth segmented region but the pixels present in the segmented region detected by the model.

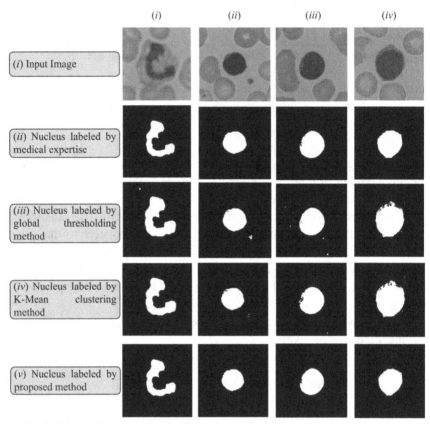

FIGURE 3.6   Nucleus segmentation by each experiment.

Jaccard similarity coefficient, also known as intersection over union (IOU) is a numeric value used to find similarities between sample sets. It is generally the ratio between the amount of intersection over the amount of union of two samples. Let I_{GT} denotes golden truth segmented area of the image and I_{US} denotes the segmented area by proposed U-Net model, then mathematical expression for finding Jaccard Index as in Eqn. (11).

$$JaccardIndex(\%) = \left[\frac{I_{GT} \cap I_{US}}{I_{GT} \cup I_{US}}\right] \times 100 \tag{11}$$

3.4.3 RESULT

The quantitative performance analysis of an arbitrary chosen 20 images of segmented nucleus by Global thresholding, the K-means, and the U-Net model for parameters such as specificity, sensitivity, Jaccard index, and accuracy. Table 3.2 shows the performance of all three discussed segmentation methods in the percentage of sensitivity and specificity parameters. Table 3.3 shows the Jaccard index and the accuracy performance of all three segmentation techniques.

TABLE 3.2 Quantitative Performance Table of All Experiments for Sensitivity and Specificity

Image No.	Sensitivity (%)			Specificity (%)		
	Experiment 1	Experiment 2	Experiment 3	Experiment 1	Experiment 2	Experiment 3
1.	94.12	95.54	84.34	94.18	97.13	100
2.	90.43	92.39	87.44	48.81	91.27	99.71
3.	90.88	91.61	92.15	97.76	94.01	98.77
4.	98.43	99.13	92.02	97.52	93.98	97.58
5.	96.13	97.22	88.07	94.63	91.81	98.64
6.	98.05	98.45	90.47	78.43	94.54	97.23
7.	90.53	90.73	94.54	58.99	95.52	95.45
8.	94.12	94.65	95.52	47.84	94.86	89.50
9.	96.25	96.55	94.86	52.40	93.74	88.81
10.	91.85	91.44	93.74	57.34	97.42	99.94
11.	91.34	92.92	97.42	60.06	95.84	96.25
12.	84.99	86.95	95.84	73.18	80.04	96.22
13.	91.98	91.67	92.11	46.26	91.17	97.88
14.	64.32	66.34	92.15	82.12	73.14	95.06
15.	84.10	87.13	93.20	62.18	84.10	95.36
16.	81.49	83.22	92.80	84.91	73.22	92.43
17.	93.11	94.03	94.77	64.99	83.27	94.85
18.	85.01	86.23	95.52	71.94	81.26	95.88
19.	87.99	88.21	97.46	63.72	83.30	96.69
20.	89.94	90.13	95.36	59.32	86.84	95.60
Average	89.753	90.727	92.989	69.829	88.823	96.0925

TABLE 3.3   Quantitative Performance Table of All Experiments for Jaccard Index and Accuracy

Image No.	Jaccard Index (%)			Accuracy (%)		
	Experiment 1	Experiment 2	Experiment 3	Experiment 1	Experiment 2	Experiment 3
1.	88.59	95.67	85.75	94.36	94.90	92.87
2.	60.07	93.15	88.12	69.85	90.79	94.04
3.	87.99	92.24	93.35	94.28	90.83	96.12
4.	94.54	99.23	93.56	98.15	97.39	96.66
5.	95.52	97.54	99.68	95.35	95.25	94.94
6.	94.86	98.59	91.38	88.27	96.61	95.69
7.	93.74	92.02	92.07	74.74	88.36	96.03
8.	97.42	95.23	88.23	71.06	92.71	93.07
9.	95.84	97.14	86.78	74.32	93.88	92.41
10.	64.20	92.95	98.54	74.57	94.54	93.14
11.	65.06	93.98	90.72	75.67	95.52	93.66
12.	66.66	88.98	95.56	79.04	94.86	94.45
13.	59.33	92.20	96.23	68.77	93.74	93.72
14.	54.72	74.07	98.14	73.68	97.42	94.74
15.	61.19	89.15	94.54	73.53	95.84	93.61
16.	70.02	87.14	95.52	82.94	80.25	93.30
17.	68.27	95.11	94.86	78.61	89.66	94.06
18.	66.53	88.80	93.74	78.72	84.09	93.99
19.	64.40	90.38	97.42	75.90	86.02	96.99
20.	63.74	91.89	95.84	74.59	88.33	95.10
Average	75.63	92.27	93.50	79.82	92.05	94.43

3.4.4   RESULT ANALYSIS

The method used in the first experiment is global thresholding for nucleus segmentation, and the performance is evaluated on the basis of values of quantitative parameters, which is shown in Tables 3.2 and 3.3. The average performance of the first method mostly ranges between 69% and 90%, which indicates the resultant nucleus given by this method does not well correlate with ground truth. The second experiment uses the K-means clustering method and performs slightly better than the prior method with a range of values 88 to 92% shown in Tables 3.2 and 3.3. It can be observed that the U-Net model for segment nucleus improved the result than the prior methods in terms of accuracy and all other parameters, and the average performance of this method ranges between 92% and 96%.

3.5 CONCLUSION

The segmentation of WBCs is a very important problem because of the irregular morphological structure of WBC nucleus, color, and size. In this work, U-Net based automatic segmentation model is proposed, which uses a sigmoid activation function with *Adam* optimization and "binary cross-entropy" loss function. These parameters are suitable for maintaining network weight with good learning rate. The experimental outputs proved that the U-Net-based method shows outstanding results in comparison to the previously published methods, including the Global thresholding and K-means technique. All the experiments highlighted that the U-net DL model is more successful technique for nucleus segmentation compared to other methods.

KEYWORDS

- acute myeloid leukemia
- complete blood cell
- convolution layer
- hidden Markov random field
- intersection over union
- morphological opening
- red-blood-cells
- white-blood-cells

REFERENCES

1. Work, T. M., Raskin, R. E., Balazs, G. H., & Whittaker, S. D., (1998). Morphologic and cytochemical characteristics of blood cells from Hawaiian green turtles. *American Journal of Veterinary Research, 59*, 1252–1257. doi: https://pubs.er.usgs.gov/publication/1003913.
2. Bain, B. J., (2008). *A Beginner's Guide to Blood Cells.* Wiley. doi: 10.1002/9781119367871.
3. Xing, F., & Yang, L., (2016). Robust nucleus/cell detection and segmentation in digital pathology and microscopy images: A comprehensive review. *IEEE Reviews in Biomedical Engineering, 9*, 234–263. doi: https://doi.org/10.1109/RBME.2016.2515127.
4. Wen, J., Xu, Y., Li, Z., Ma, Z., & Xu, Y., (2018). Inter-class sparsity based discriminative least square regression. *Neural Networks, 102*, 36–47. doi: https://doi.org/10.1016/j.neunet.2018.02.002.
5. Bhatt, C., Kumar, I., Vijayakumar, V., Singh, K. U., & Kumar, A., (2020). The state of the art of deep learning models in medical science and their challenges. *Multimedia Systems*, 1–15. doi: https://doi.org/10.1007/s00530-020-00694-1.
6. Mandyartha, E. P., Anggraeny, F. T., Muttaqin, F., & Akbar, F. A., (2020). *Global and Adaptive Thresholding Technique for White Blood Cell Image Segmentation in Journal of Physics: Conference Series* (Vol. 1569, No. 2, p. 022054). IOP Publishing. DOI: https://iopscience.iop.org/article/10.1088/1742-6596/1569/2/022054.
7. Zhou, X., Li, Z., Xie, H., Feng, T., Lu, Y., Wang, C., & Chen, R., (2020). Leukocyte image segmentation based on adaptive histogram thresholding and contour detection. *Current Bioinformatics, 15*(3), 187–195. doi: https://doi.org/10.2174/1574893614666190723115832.
8. Ostu, N., (1979). A threshold selection method from gray-level histograms. *IEEE Transactions on Systems, Man, and Cybernetics, 9*(1), 62–66. doi: https://doi.org/10.1109/TSMC.1979.4310076.
9. Zhang, C., Xiao, X., Li, X., Chen, Y. J., Zhen, W., Chang, J., & Liu, Z., (2014). White blood cell segmentation by color-space-based k-means clustering. *Sensors, 14*(9), 16128–16147. doi: https://doi.org/10.3390/s140916128.
10. Su, J., Liu, S., & Song, J., (2017). A segmentation method based on HMRF for the aided diagnosis of acute myeloid leukemia. *Computer Methods and Programs in Biomedicine, 152*, 115–123. doi: https://doi.org/10.1016/j.cmpb.2017.09.011.
11. Inbarani, H. H., & Azar, A. T., (2020). Leukemia image segmentation using a hybrid histogram-based soft covering rough k-means clustering algorithm. *Electronics, 9*(1), 188. doi: https://doi.org/10.3390/electronics9010188.
12. Kumar, I., Bhatt, C., & Singh, K. U., (2020). Entropy based automatic unsupervised brain intracranial hemorrhage segmentation using CT images. *Journal of King Saud University-Computer and Information Sciences.* doi: https://doi.org/10.1016/j.jksuci.2020.01.003.
13. Drozdzal, M., Vorontsov, E., Chartrand, G., Kadoury, S., & Pal, C., (2016). The importance of skip connections in biomedical image segmentation. In: *Deep Learning*

and *Data Labeling for Medical Applications* (pp. 179–187). Springer, Cham. doi: https://doi.org/10.1007/978-3-319-46976-8_19.

14. LeCun, Y., Bengio, Y., & Hinton, G., (2015). Deep learning. *Nature, 34*(5), 521. doi: https://doi.org/10.1038/nature14539.

15. Labati, R. D., Piuri, V., & Scotti, F., (2011). All-IDB: The acute lymphoblastic leukemia image database for image processing. In: *2011 18th IEEE International Conference on Image Processing* (pp. 2045–2048). IEEE. doi: https://doi.org/10.1109/ICIP.2011.6115881.

16. Duan, T., Tang, Y., Gao, F., & Yao, J., (2019). Application of mathematical morphological filter for noise reduction in photoacoustic imaging. In: *Photons Plus Ultrasound: Imaging and Sensing 2019* (Vol. 10878, p. 1087854). International Society for Optics and Photonics. doi: https://doi.org/10.1117/12.2512176.

17. Hajian-Tilaki, K., (2013). Receiver operating characteristic (ROC) curve analysis for medical diagnostic test evaluation. *Caspian Journal of Internal Medicine, 4*(2), 627. doi: https://www.ncbi.nlm.nih.gov/pmc/articles/PMC3755824.

CHAPTER 4

Introduction and Application of SVM in Brain Tumor Segmentation

AMIT VERMA

School of Computer Science, UPES, Dehradun, Uttarakhand, India

ABSTRACT

Automate the process of brain tumor segmentation using the concepts of machine learning and deep learning (DL) is one of the most emerging research topics nowadays. Manually segmenting the malignant area in the brain is tedious work which is done by a radiologist to prepare a report declaring the size of a tumor in the brain. The operator sees the MR image of a patient and uses some graphical tools to segment the malignant area in the MRI of the brain, based on his/her experience. The doctor further diagnoses the patient based on the report. Accurately automating this process is one of the challenging tasks, which involves the process of outlining the malignant part of the brain in MR image. The large difference between the population of normal and malignant tissues and the highly diverse geometry of the tumors makes it more difficult to build an accurate hypothesis model for segmenting the brain tumor. Multiple segmentation techniques are used in brain tumor segmentation such as Support Vector Machine (SVM), Convolutional Neural Network, K-mean clustering, Hidden Markov Model (HMM), random walk, etc. In this chapter, a popular classification approach SVM is discussed with its application in segmenting the brain tumor. Applying SVM on selected images of brain tumor from BRATS 2013 data set give 94%, 93%, and 96% of dice score, sensitivity, and positive protected value (PPV) for complete tumor in MR image (Table 4.1).

Application of Deep Learning Methods in Healthcare and Medical Science.
Rohit Tanwar, PhD, Prashant Kumar, PhD, Malay Kumar, PhD, & Neha Nandal, PhD (Editors)
© 2023 Apple Academic Press, Inc. Co-published with CRC Press (Taylor & Francis)

4.1 INTRODUCTION

According to the American society ASCO [1] nowadays death rate due to cancer is spreading rapidly. The brain tumor is among the leading cancer nowadays, which require surgical operation after the detection of the tumor in the brain [2]. A brain tumor can be broadly categorized into two types is Benign and Malignant. Benign tumors don't affect the neighbor tissues and remain stagnant in size for a long time, whereas Malignant tumors grow rapidly and effecting the neighbor tissues also to increase the tumor size aggressively [3]. To study the anatomy of the brain, Medical Resonance Imaging (MRI) technique is proved to be a great achievement. MR imaging is used to see the detailed anatomy of the brain, which helps the operator to segment the tumor structure that is the Region of Interest (ROI) in the brain so, that the doctor can effectively diagnose the tumor patient. Still, multiple latent features could help doctors better understand the tumor condition of the patient and cannot be identified manually by the operators. There are multiple Computer-Aided Diagnosis (CAD) systems are designed to solve the purpose of detecting the size, shape, and location of the brain tumor [4]. Due to the complexity of the MR images, automating the task of segmenting the ROI is a tedious task but is an area of interest for many researchers. Many techniques for brain tumor segmentation were proposed based on some state-of-arts classification techniques, used for this purpose such as SVM, CNN, HMM, and many more.

Image segmentation [5–9] is a process of exploring the area of interest in the image. It plays a very important role in clinical image analysis. In image segmentation, the image is subdivided into various parts, and performing the analysis of the ROI in the image based on some voxel features. In the case of brain tumor segmentation also, based on a dataset of MR images the model is trained to classify the malignant region of the brain in the MR image. Multiple image segmentation techniques are applied for segmenting the clinical MR images. Region-based segmentation method [10], like region growing approach in which starting from initial seed pixels, neighborhood pixels are added with the seeds based on some similar properties. Thresholding-based segmentation [11], the image is segmented by converting the gray image into a binary image that is black and white based on the intensity of each pixel of the original image. Various learning-based approaches are also used for segmenting the MR images by classifying each pixel of the image to get the ROI. In this chapter, we will study one state-of-art method of classification that is SVM, and its application in brain tumor segmentation.

4.2   APPLICATION OF SVM IN BRAIN TUMOR SEGMENTATION

SVM technique can be used for both classification and regression problems [12], as here we are using SVM for clinical image segmentation therefore, we consider the classification approach. SVM is one of the popular classification techniques as it is having a clever way of preventing overfitting and with comparative less computation classification can be done based on a large number of features or multi-dimensional data. The problem of overfitting [13] as shown in Figure 4.1 can be solved using SVM. As shown in the Figure 4.1 representing the data with red and black dots belonging to two different classes. The blue line, which is representing the decision boundary is perfectly classifying the data. This condition gives rise to the problem of overfitting in which the model fails to fit any additional data and predicting results for any input. Overfitting is more likely to happen with nonlinear data. Further, we will discuss an introduction to SVM and its application in brain tumor segmentation in this topic.

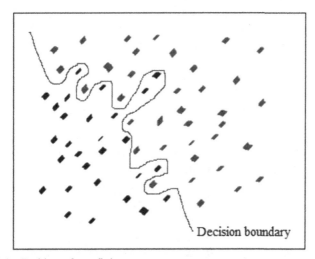

Decision boundary

FIGURE 4.1   Problem of overfitting.

4.2.1   INTRODUCTION TO SVM

SVM is a quite popular classification technique [14], apart from that there are multiple classification techniques to classify the data according to the features. Such a logistic regression (LR) is one way for partitioning the data [15] or more preferable non-linear data into two or multiple classes, as shown in Figure 4.2, where we have data of two different classes. The decision boundary is dividing the data into two separate classes for predicting the unclassified input [16]. LR uses sigmoid or logistic function as shown in Eqn. (1) to draw a decision boundary for the classification of data based on some features.

$$g(z) = \frac{1}{1 + e^{-z}}$$

(1)

This sigmoid function can be graphically represented as shown in Figure 4.2.

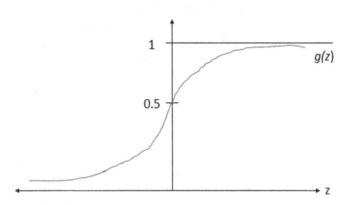

FIGURE 4.2   Sigmoid function.

LR model that is the hypothesis function based on LR can be represented by Eqns. (2) and (3) as shown below:

$$h_\theta(x) = g(\theta^T x)$$

(2)

$$h_\theta(x) = \frac{1}{1 + e^{-\theta^T x}}$$

(3)

According to Eqn. (4) if the probability that $Y = 1$, for given x, parameterized by is greater than 0.5, then the new input will be in negative class (0) according to Figure 4.2 else x will belong to a positive class that is 1.

$$h_\theta(x) = P(Y = 1|x, \theta)$$

(4)

LR use for data with fewer features as shown in Figure 4.3 that is it is not very suitable for classifying the data with the higher number of features. In Figure 4.3 let the y-axis and x-axis are representing two different features x_1 and x_2, red points are negative points, and black are positive points. The blue line is representing the decision boundary. Now for any new data, if the output is close to the decision boundary, then we would have less confidence as compared to the output away from the decision surface (DS). If we are having a large number of features, then Bayesian learning (BL) [17] can be the better option, but BL is computationally unattractive.

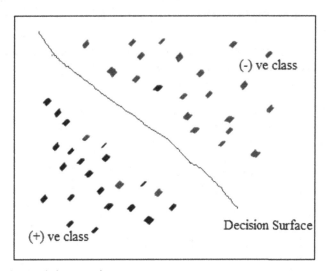

FIGURE 4.3   Logistic regression.

SVM is a good alternative for BL, which is computationally attractive, in SVM we want a classifier that maximizes the distance of negative and positive points from the DS [18]. In SVM, we find the distance of each negative and positive point from the DS. The distance of the point closest to the DS is considered as minimum margin M1 as shown in Figure 4.4(a) with the bold blue line between the closest positive point and the DS. Similarly, we would have a margin of M2 from the negative point that is the distance between the negative point closest to the DS as shown in Figure 4.4(b). Now, if we shift the DS in such a way that the margin M1 from positive point and margin M2 from negative point become approximately equal as shown in Figure 4.5.

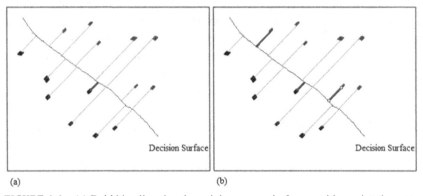

(a)   (b)

FIGURE 4.4   (a) Bold blue line showing minimum margin from positive point closest to DS; (b) distance of two positive points from DS.

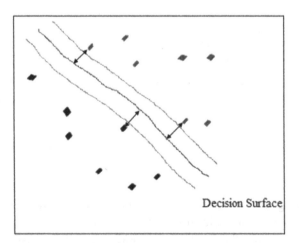

FIGURE 4.5   Shifted DS and M1 and M2 are approximately same.

As shown in Figure 4.5, the DS is shifted in such a way that the margins M1 and M2 are approximately equal and the width of the path represented by two green lines in Figure 4.5 is maximum in this case. In the same way, there could be multiple paths according to the given points therefore the main objective of SVM is to find the best suitable DS with the maximum path width to classify the data.

4.2.2   BRAIN TUMOR SEGMENTATION USING SVM

Multiple researchers have contributed to the field of brain tumor segmentation that is outlining the region of the brain with malignant tissues using SVM. Before applying the actual segmentation method, the brain MR images are preprocessed, to enhance the images and remove noise for better results. Major steps of pre-processing of the images are acquisition, reconstruction, co-registration, template registration, extraction of a slice of interest, and noise reduction [19]. Acquisition of MR images is the first and the important step of BTS as the results of the remaining process depend on the quality of the images acquired. A wide variety of sensors are also used for this step [20]. After this step, we get data set of images (2D slices). The acquisition is followed by reconstruction, in this step, 2D images are converted into 3D images for better analysis of the BT and to provide a clearer view for diagnosis by the doctor. This conversion can be done using software like XMedCon [21]. And co-registration can be done using SPM software which is followed by template registration (SPM software). Now we extract the slices of interest to avoid the rest of the image part, to be focused on the tumor region, and remove the redundant area. The final step is noise reduction, MRIs suffer from various noises such as Gaussian, speckle, and salt and pepper noise [22]. Noise in simple words can be defined as the variation or deviation in contrast or brightness level of the images. Gaussian noise occurs during the acquisition of images which can be due to sensor noise/electronic circuit noise caused by variation in the temperature/transmission. Salt and pepper noise is impulse noise, salt noise is having white pixels on the black background and pepper noise is just the opposite of the salt noise. A final step that is noise reduction, requires the smoothening of the images by using various filters such as a spatial filter for removing noise. After pre-processing has been done, feature extraction is carried out. The complete procedure of segmenting the brain tumor in MR images using SVM is shown in Figure 4.6.

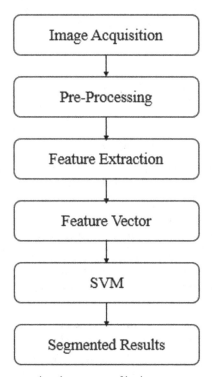

FIGURE 4.6   Steps representing the process of brain tumor segmentation.

1. **Feature Extraction:** Pixel features play an important role in the classification of the pixel either as healthy or malignant tissue. By the term features of a pixel is the properties of the pixel, the main features of the pixel are texture and intensity. Both the properties are important for the classification of the pixel [23, 24]. These are the common use features in BTS [25]. The texture and intensity level of each pixel in the MR images can be used to categorize the pixel in some modality T1 and T2. There are multiple ways of feature extraction for the classification of the pixels in MRIs, further, the pixel texture could be categorized as first-order and second-order texture. It is important to generate the relevant

features for better classification. Feature generated from the FOT parameter provides the information about the distribution of gray level in the image but does not provide any information about the position of the gray-level. The information about the position and direction of the gray-level in the image are gathered by SOT parameters [26].

2. **Training and Testing:** In this phase, the dataset is distributed or divided into training and testing data sets. The data sets have data in form of MR images with ground truth so, data is divided into some ratios, for example, 7:3 is 70% data to train the model and 30% data for testing. With the 70% data, the model is trained using the SVM technique as discussed in Section 4.2.1. After training the model, testing data is used to test the model for accuracy. Experimental results of applying SVM on five selected MR images of brain tumor is shown in Figure 4.7. Applying SVM on the MR images of brain tumor for complete and enhancing tumor give results as shown in Table 4.1. One of the major problems with the model trained on data sets of medical images is the biasness of the model towards the positive classification that is healthy tissue [27]. This is happening due to the large population of pixels representing the healthy tissues in MR images. The problem can be rectified using bagging or boosting techniques [28].

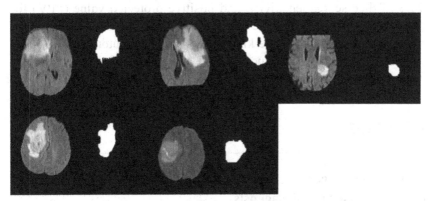

FIGURE 4.7   Results are shown for 5 MR images of brain tumor segmenting the complete tumor.

TABLE 4.1   Accuracy Results of SVM on MR Images for Complete and Enhancing Tumor

Model	Dice Score		Sensitivity		PPV	
	Complete	Enhanc-ing	Complete	Enhanc-ing	Complete	Enhanc-ing
SVM on original data	0.9494	0.7096	0.9373	0.7773	0.9618	0.6528

4.3 CONCLUSION

In this chapter, we discussed the importance of automation in the field of brain tumor segmentation based on MR images so that the doctor can conduct the better diagnosis and report can be uniform in different MR imaging centers. For performing segmentation of brain tumor that is separating the healthy and the tumor area in the MR image, there are multiple techniques in which SVM (Support Vector Machine) is one of the popular techniques. Considering the same, in this chapter detail introduction of SVM is carried out with the required figures for better understanding. The problem of overfitting is also highlighted. All the necessary steps of performing MR image segmentation are explained followed by training and testing of the model based on the brain tumor data set. Applying SVM for BTS gives the accuracy as 94%, 93%, and 96% of dice score, sensitivity, and positive protected value (PPV) for complete tumor in MR image as shown in Table 4.1. By the reading of this chapter, viewers get the knowledge about the automation of brain tumor segmentation, introduction, and application of SVM, and steps required for segmentation.

KEYWORDS

- Bayesian learning
- computer-aided diagnosis
- hidden Markov model

- **logistic regression**
- **medical resonance imaging**
- **positive protected value**
- **region of interest**
- **support vector machine**

REFERENCES

1. American Society of Clinical Oncology (ASCO). http://www.asco.org/ (accessed on 14 May 2022).
2. Huo, J., Brown, M. S., & Okada, K., (2012). CADrx for GBM brain tumors: Predicting treatment response from changes in diffusion-weighted MRI. In: *Machine Learning in Computer-Aided Diagnosis: Medical Imaging Intelligence and Analysis* (pp. 297–314). IGI Global.
3. John, P., (2012). Brain tumor classification using wavelet and texture based neural network. *International Journal of Scientific & Engineering Research, 3*(10), 1–7.
4. Dandıl, E., Çakıroğlu, M., & Ekşi, Z., (2014). Computer-aided diagnosis of malign and benign brain tumors on MR images. In: *International Conference on ICT Innovations* (pp. 157–166). Springer, Cham.
5. Hamamci, A., Kucuk, N., Karaman, K., Engin, K., & Unal, G., (2012). Tumor-cut: Segmentation of brain tumors on contrast-enhanced MR images for radiosurgery applications. *IEEE Transactions on Medical Imaging, 31*(3), 790–804.
6. Havaei, M., Larochelle, H., Poulin, P., & Jodoin, P. M., (2016). Within-brain classification for brain tumor segmentation. *International Journal of Computer Assisted Radiology and Surgery, 11*(5), 777–788.
7. Tustison, N. J., Shrinidhi, K., Wintermark, M., Durst, C. R., Kandel, B. M., Gee, J. C., Grossman, M. C., & Avants, B. B., (2015). Optimal symmetric multimodal templates and concatenated random forests for supervised brain tumor segmentation (simplified) with ANTsR. *Neuroinformatics, 13*(2), 209–225.
8. Bal, A., & Saha, R., (2016). An improved method for handwritten document analysis using segmentation, baseline recognition and writing pressure detection. *Procedia Computer Science, 93*, 403–415.
9. Kwon, D., Shinohara, R. T., Akbari, H., & Davatzikos, C., (2014). Combining generative models for multifocal glioma segmentation and registration. In: *International Conference on Medical Image Computing and Computer-Assisted Intervention* (pp. 763–770). Springer.
10. Shanthi, K., & Kumar, M. S., (2007). Skull stripping and automatic segmentation of brain MRI using seed growth and threshold techniques. In: *International Conference on Intelligent and Advanced Systems, 2007 (ICIAS 2007)* (pp. 422–426). IEEE.
11. Taheri, S., Ong, S. H., & Chong, V., (2010). Level-set segmentation of brain tumors using a threshold-based speed function. *Image and Vision Computing, 28*(1), 26–37.

12. Gunn, S. R., (1998). Support vector machines for classification and regression. *ISIS Technical Report, 14*(1), 5–16.
13. Dietterich, T., (1995). Overfitting and under computing in machine learning. *ACM Computing Surveys (CSUR), 27*(3), 326, 327.
14. Mathur, A., & Giles, M. F., (2008). Multiclass and binary SVM classification: Implications for training and classification users. *IEEE Geoscience and Remote Sensing Letters, 5*(2), 241–245.
15. Fienberg, S. E., Nardi, Y., & Slavković, A. B., (2008). Valid statistical analysis for logistic regression with multiple sources. In: *Annual Workshop on Information Privacy and National Security* (pp. 82–94). Springer, Berlin, Heidelberg.
16. Wang, D., Zhang, M., Li, Z., Cui, Y., Liu, J., Yang, Y., & Wang, H., (2015). Nonlinear decision boundary created by a machine learning-based classifier to mitigate nonlinear phase noise. In: *2015 European Conference on Optical Communication (ECOC)* (pp. 1–3). IEEE.
17. Ch, R., & Map, M., (1997). Bayesian learning. *Book: Machine Learning* (pp. 154–200). McGraw-Hill Science/Engineering/Math.
18. Pontil, M., & Verri, A., (1998). Properties of support vector machines. *Neural Computation, 10*(4), 955–974.
19. Ayachi, R., & Amor, N. B., (2009). Brain tumor segmentation using support vector machines. In: *European Conference on Symbolic and Quantitative Approaches to Reasoning and Uncertainty* (pp. 736–747). Springer, Berlin, Heidelberg.
20. Rajasekaran, K. A., & Gounder, C. C., (2018). Advanced brain tumor segmentation from MRI images. *High-Resolution Neuroimaging: Basic Physical Principles and Clinical Applications,* p. 83.
21. X Medical Conversion. Available at https://xmedcon.sourceforge.io/ (accessed on 14 May 2022).
22. Ali, H. M., (2018). MRI medical image denoising by fundamental filters. *High-Resolution Neuroimaging-Basic Physical Principles and Clinical Applications,* 111–124.
23. Dickson, S., & Thomas, B., (1997). Using neural networks to automatically detect brain tumors in MR images. *International Journal of Neural Systems, 4*(1), 91–99.
24. Tuceryan, M., & Jain, A., (1998). *Texture Analysis* (2nd edn., pp. 207–248). World Scientific Publishing, Singapore.
25. Schmidt, M., (2005). *Automatic Brain Tumor Segmentation.* University of Alberta, Department of computing science.
26. Aggarwal, N., & Agrawal, R. K., (2012). *First and Second-Order Statistics Features for Classification of Magnetic Resonance Brain Images. 3*(2), 146–153.
27. Zacharaki, E. I., Wang, S., Chawla, S., Soo, Y. D., Wolf, R., Melhem, E. R., & Davatzikos, C., (2009). Classification of brain tumor type and grade using MRI texture and shape in a machine learning scheme. *Magnetic Resonance in Medicine: An Official Journal of the International Society for Magnetic Resonance in Medicine, 62*(6), 1609–1618.
28. Kotsiantis, S., & Pintelas, P., (2004). Combining bagging and boosting. *International Journal of Computational Intelligence, 1*(4), 324–333.

CHAPTER 5

Detection Analysis of COVID-19 Infection Using the Merits of Lungs CT Scan Images with Pre-Trained VGG-16 and 3-Layer CNN Models

P. VIJAYALAKSHMI,[1] P. SATHISH KUMAR,[2] and V. RAJENDRAN[1]

[1]Department of Electronics and Communication Engineering, Vels Institute of Science, Technology, and Advanced Studies (VISTAS), Chennai, Tamil Nadu, India, E-mail: viji.se@velsuniv.ac.in (P. Vijayalakshmi)

[2]Department of Electronics and Communication Engineering, Bharath Institute of Higher Education and Research, Selaiyur, Chennai-73, Tamil Nadu, India, E-mail: sathishmrl30@gmail.com

ABSTRACT

Coronavirus disease is a cruel worldwide issue for which artificial intelligence (AI) technique, especially deep learning (DL)-based CNN approach, plays a significant role in reduction in death rate by physical presence of examination on lungs in an earlier stage. In our proposed work, we introduced automatic recognition of infectivity in lungs from computer-aided tomography (CAT) images provide a huge prospective to enhance the conventional healthcare stratagem for dealing with coronavirus disease. Moreover, the images are segmented to point out the infected region in the lung on CAT scan images for distinguishing normal images from COVID-19 patients CT scan images. Among various AI methods,

Application of Deep Learning Methods in Healthcare and Medical Science.
Rohit Tanwar, PhD, Prashant Kumar, PhD, Malay Kumar, PhD, & Neha Nandal, PhD (Editors)
© 2023 Apple Academic Press, Inc. Co-published with CRC Press (Taylor & Francis)

the DL-based VGG-16 and 3 layer CNN model attains better accuracy in finding and segmenting the images by evaluating metrics such as loss, accuracy, validation loss and validation accuracy.

5.1 INTRODUCTION

Coronavirus disease is a deadly dangerous disease that initially emerged in Wuhan City, China at the end of year 2019 December. This syndrome is caused by SARS-Corona Virus 2, a type of virus which belongs to huge family unit of coronaviruses. The major signs of coronaviruses are dry cough, uncontrolled fever, as well as drowsiness. Moreover, other symptoms such as stains, throbbing, and also complications while inhalation may happen in some other patients also. The majority of signs are the evidence for inhalation infectivity along with respiring defect that can be diagnosed by medical doctors who are familiar in finding disease as well as damages using medicinal images and taking care of the disease.

Nowadays, recent technologies like machine learning and deep learning (DL) methods easily detect COVID-19 disease within 5 minutes. Accordingly, it is probable to utilize predictive analytics (ML) algorithms to discover the disease from medical images such as computer-aided tomography (CAT) scan images, mammogram (X-ray) images, and ultrasound (US) images. Automatic appliances know how it assists medical doctors who detect the disease and injury. Hence, we are introducing novel DL algorithm namely VGG-16, and three-layer CNN for both raw images and also segmented images that has the capacity to detect the coronavirus disease automatically.

5.2 PROBLEM DESCRIPTION

The major issues related with coronavirus disease are described below:

- When compared to mammogram (X-ray) images of a particular organ, i.e., here the specified organ as lung, computed tomography (CT) viewing is broadly chosen because of its improvement and 3-D vision of lung.

- In advanced studies, the typical signs of infection could be observed from CT scans, for instance finding pulmonary consolidation in later stage and ground-glass opacity in earlier stages.

To resolve the above issues, we are introducing our novel proposed CNN model namely VGG-16, three-layer CNN to segment the lung infected images by finding classification accuracy, validation accuracy and loss.

The main objective of this proposed work is:

- to build image classifiers to determine whether the patient is tested positive or negative for coronavirus disease from lung CT scan images;
- to develop DL-based VGG-16 model as well as three-layer CNN model for classifying CT images;
- to build machine learning model, namely K-means clustering for segmenting CT image.

5.3 BACKGROUND

Afsin et al. [1] reviewed many research work regarding how the DL techniques are applicable in medical domains as coronavirus disease prediction using CT images and mammogram. In addition to that, we survey many papers focused on disease occurrence in various locations via DL methods. Finally, the confront appearance in self-recognition of Coronavirus using deep structured learning were conversed. Ahmed et al. [2] For coronavirus disease analysis, the researcher introduced novel detection approach namely end-to-end semi-supervised algorithm in which ResNext+ which have need of volume level data and supply slice level prediction. To mine the spatial features, integration of lung segmentation mask and also spatial and direct attention were added in. Finally, LSTM (Long Short Term Memory) were used to attain axial dependency of very region of images. The precision and F1-score metrics were estimated is shown in Table 5.1 [3].

TABLE 5.1   Precision and F1-Scores

Datasets	Precision	F1-Score
Public available dataset	81.9%	81.4%
Closest state of art	76.7%	78.8%

From the above work, they conclude that the precision metrics enhancement is nearly 5% and F1-score is 3% that expressed the proposed methodology efficiency. Amine Amyar et al. [4] introduced a novel technique (multi-tasks DL-based model) comprised of three learning tasks namely segmenting images, classifying images and finally modernizing the images are mutually performed with various datasets. It has single encoder for parameter illustration and two decoders along with multilayer perceptron (MLP) for executing multitasking model. Michael et al. [5] exhibited how the DL model in particular transfer learning method may be utilized in COVID-19 disease from CT, mammogram, and US images. Moreover, the researcher analyzed VGG-19 DL image classification model to diagnose the COVID-19 disease. Through this analysis, the outcomes signify that US images afford better-quality detection accuracy when compared to mammogram images and CT scans. The accuracy detection for all three images as US images (100%), Mammogram images (86%), and CT images (84%) based on precision. Mohammed et al. [6] discovered rapid as well as precise fully automated technique to identify coronavirus disease patient from CT images. Initially, the investigator applied image processing method to find disease but the lungs image is not clearly seen in CAT scan. Hence, they introduced CNN based model namely Residual Network 50 Version 2 for further implementation to categorize the input CAT images as non-COVID individuals and COVID-19 patients in Table 5.2 [6].

TABLE 5.2   Accuracy Prediction

CT Images	Accuracy
Tested images 7,996	98.5%

Among these tested images, 234 images were correctly classified along with a very high rate. Moreover, the researcher examined the infected region on lung CT images using Gradient Camera algorithm able to envisage the layers of CNN and also calculate the model classification exactness.

Parisa et al. [7] constructed ensemble approach derived from higher voting of preeminent amalgamation of DL algorithms for improving model detection performance. For generating better performance, datasets related to lung CT images have gathered and finally labeled as COVID-19 (positive) and Normal (Negative) images. The metrics like precision, recall, and accuracy were found to predict the Coronavirus disease from CT images. Pedro et al. [8] introduced the DL method especially voting-based approach for screening coronavirus patients easily. Here the collected images are categorized as grouping based on a voting system. Using this approach, the investigator trialed in two large datasets for CT image analyzes have done depends on patient split which is helpful for medicinal fields as a real time circumstances.

Deng et al. [9] introduced the Inf-Network technique which is a Lung infection Segmentation DL method that helps to divide the infected region in lungs from CT images without human intervention. Therefore, this approach is useful in finding normal individual and Coronavirus disease affected patients easily using CT images. Also, semi-supervised learning algorithm enhances the learning capability as well as attains greater performance. Halgurd et al. [10] developed a novel AI tool for diagnosing disease like coronavirus which are supportive in medical fields, especially in physical condition care expertise. The major demanding task of using this tool is whether it is applicable in case of being deficient datasets related to CT and mammogram images. Also, CNN models like transfer learning, AlexNet model were applied on CT and mammogram images. The accuracy estimated using two DL models are mentioned in Table 5.3.

TABLE 5.3 The Model Used to Predict COVID-19 Along with Accuracy Estimation by Maghdid et al. [10]

Model Used	Accuracy
AlexNet	98%
Modified CNN	94.1

Arnab et al. [11] suggested a decision fusion based method that comprises of many separate methods to generate ultimate diagnosis of disease on CAT images. This approach will be useful in detecting coronavirus disease very earlier. The metrics found as percentage by Arnab et al. [11] is mentioned in Table 5.4.

TABLE 5.4   Metrics with Percentage by Arnab et al. [11]

Metrics	Percentage
Accuracy	86%
AUROC	88%
F1-score	87%

Jamshidi et al. [12] introduced AI, especially DL algorithms such as LSTM (Long Short Term Memory), Machine learning extreme algorithms, and some AI-based networks to diagnose coronavirus disease.

Sadman et al. [13] recommended a feasible and effective DL classification algorithm based on chest radiography (DL-CRC) structure that classifies the abnormal (pneumonitis) and abnormal (coronavirus disease) from normal individuals. Ilker et al. [14] reviewed several articles around 130 COVID research papers to compare which AI technique is suitable for identifying the disease from US images, Mammogram images, and Chest CAT images. Metrics such as sensitivity, recall, accuracy, F measure and ROC were evaluated by comparing different studies to find the model performance.

Sweta et al. [15] surveyed how AI techniques especially DL approach helpful in medicinal healthcare domain in detecting all kind of disease as well as distinguishing normal from abnormal patients on either US images, mammogram, and CT scan images.

Jianpeng et al. [16] built a novel DL based abnormality findings approach for high speed, trustworthy screening. The datasets gathered for this work is 100 mammogram images on chest, For DL model these images are very least hence they collected more images around 1,431 chest mammogram images to perform screening. Adil et al. [17] designed an integration of CNN and RNN specified as Deep Sense method. This approach is intended as a sequence of several layers to mine and distinguish the relevant parameters of coronavirus disease from lungs.

Deepika et al. [18] developed CNN model to detect pneumonitis disease from chest mammogram images based on feature extraction carried out on chest images and finally classification of images as normal and pneumonitis. Harsh et al. [19]; and Yifan et al. [20]; Tao et al. [21] disease identification and diagnosis have done via RT-PCR testing having greater accuracy. Also, this investigator applied DL based transfer learning algorithm on mammogram and CAT images to diagnose the COVID-19.

Kai et al. [22] various CAT parameters have seen in accordance with severity of disease that makes helpful in finding COVID-19 disease stratification. Maayan et al. [23] examined the novel fully CNN framework for performing segmentation of images such as heart, lungs, chest in several classes. Moreover, investigation had done in authority of utilizing several failure tasks (loss functions) during the training process on NN for segmenting images. Rajinikanth et al. [24]. Machine learning systems have been proposed to detect patients infected with COVID-19 via radiological imaging. Also, Harmony-search, and Otsu-based system for coronavirus disease (COVID-19) detection using lung CT scan images.

5.4 EXPERIMENTAL METHODS

The experimentation aspects along with the step-by-step valuation to test out the proposed model efficiency are described in this section. Here, our demonstration is mainly rooted on Coronavirus Lung infection segmentation using CT images. The implementation have completed using Python programming language, especially in version python 3.7 along with open CV libraries since it sustains many programming languages like C, C++, Python, and so on. Moreover, we installed Keras library as non-proprietary DL structure along with Tensor Flow as a backend to construct CNN model and also perform model training. The experimentation have fully worked on standard personal computer having Intel core i5 8th generation.

The following are the modules implemented to work out how the diseases were detected on CT scan images as well as image classification were performed:

* **Module 1:** Dataset preparation.
* **Module 2:** Data Pre-processing.
* **Module 3:** Classification using VGG-16 model.
* **Module 4:** Classification using simple 3-layer CNN model.

The modules for segmenting the input CAT images into normal and COVID-19 are illustrated in Figure 5.1.

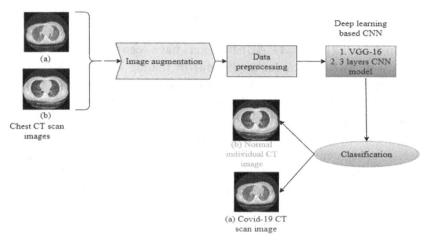

FIGURE 5.1   Proposed modules to predict and classify the images.

5.4.1   MODULE 1: DATASET PREPARATION

In this investigation, the dataset utilized for our work are listed in Table 5.5 to detect the disease along with distinguishing as normal from abnormal though CT images.

TABLE 5.5   Dataset Description

COVID-19 CT segmentation CT image	https://medicalsegmentation.com/covid19/
COVID-19 CT collection CT image	https://arxiv.org/abs/2003.11597
COVID-CT-dataset CT image	https://arxiv.org/abs/2003.13865

5.4.2   MODULE 2: DATA PRE-PROCESSING

We applied preprocessing technique for removing irrelevant features from the CT images [24] that makes greater enhancement in images and also quality of images is improved. Hence, the image segmentation has been performed for identifying and classifying the coronavirus patients CT images from normal individuals.

The image segmentation of COVID-19 positive images are depicted in Figure 5.2(a) and COVID-19 negative images in Figure 5.2(b) where the left side image represents the initial CT images and the right-side images specifies the segmented images to point out the lung which is infected.

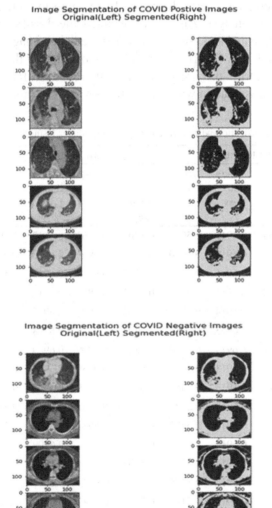

FIGURE 5.2   Image segmentation of COVID positive images (a); and negative images (b).

5.4.3   *MODULE 3: CLASSIFICATION USING VGG-16 MODEL*

After pre-processing, the next stage is to segment the images to mention the pixels and to crop the Chest CT images. This permits to spotlight on area of images for future analysis. Also, we have undergone classification of images with both segmented and raw images using medicinal image segmentation techniques such as VGG-16 model and classification of three-layer CNN model to categorize the images as normal from COVID-19 from CAT images. The classification with segmented is shown in Tables 5.6 and 5.7.

TABLE 5.6   Classification with Segmented and Raw Images Using VGG-16 Model

	Segmented Images				Raw Images			
Epoch	Loss	Validation Loss	Accuracy	Validation Accuracy	Loss	Validation Loss	Accuracy	Validation Accuracy
1/20	0.7234	0.6915	0.5026	0.5302	0.6963	0.6917	0.4902	0.5278
2/20	0.6890	0.6919	0.5620	0.5302	0.6902	0.6923	0.5485	0.5208
3/20	0.6927	0.6916	0.5374	0.5302	0.6912	0.6910	0.5496	0.5417
4/20	0.6922	0.6914	0.5247	0.5302	0.6897	0.6917	0.5608	0.5278
5/20	0.6937	0.6914	0.5090	0.5302	0.6894	0.6917	0.5572	0.5278
6/20	0.6935	0.6913	0.5178	0.5302	0.6918	0.6918	0.5296	0.5378
7/20	0.6911	0.6914	0.5330	0.5302	0.6899	0.6909	0.5552	0.5247
8/20	0.6947	0.6913	0.5069	0.5302	0.6939	0.6916	0.5072	0.5278
9/20	0.6911	0.6913	0.5357	0.5302	0.6879	0.6908	0.5645	0.5347
10/20	0.6856	0.6913	0.5784	0.5302	0.6930	0.6910	0.5170	0.5347
11/20	0.6903	0.6914	0.5427	0.5302	0.6911	0.6907	0.5320	0.5347
12/20	0.6881	0.6914	0.5676	0.5302	0.6917	0.6909	0.5311	0.5347
13/20	0.6907	0.6913	0.5367	0.5302	0.6908	0.6909	0.5389	0.5347
14/20	0.6917	0.6914	0.5296	0.5302	0.6914	0.6897	0.5347	0.5386
15/20	0.6927	0.6914	0.5184	0.5302	0.6917	0.6908	0.5292	0.5347
16/20	0.6915	0.6913	0.5302	0.5302	0.6912	0.6916	0.5366	0.5378
17/20	0.6894	0.6913	0.5506	0.5302	0.6953	0.6909	0.5021	0.5347
18/20	0.6859	0.6913	0.5780	0.5302	0.6963	0.6917	0.4902	0.5278
19/20	0.6933	0.6916	0.5184	0.5302	0.6902	0.6923	0.5485	0.5208
20/20	0.6914	0.6914	0.5341	0.5302	0.6912	0.6910	0.5496	0.5417

5.4.4 MODULE 4: CLASSIFICATION USING THREE LAYER CNN MODEL

Also, we introduced three layer CNN classification model to segment the images into pixels. Based on pixels on images, we can easily find the lung infected region in CT images. Then image segmentation process has applied to separate the COVID patient's images from normal images. The three-layer model CNN model applied on both segmented images as well as raw images for predicting training loss, validation loss, training accuracy and validation accuracy till 20 epochs by tuning hyperparameters.

TABLE 5.7 Classification with Segmented and Raw Images Using Three Layer CNN Model

	Segmented Images				Raw Images			
Epoch	Loss	Validation Loss	Accuracy	Validation Accuracy	Loss	Validation Loss	Accuracy	Validation Accuracy
1/20	0.7520	0.6522	0.5491	0.5973	1.2113	0.6265	0.4979	0.5694
2/20	0.5664	0.6127	0.7379	0.6242	0.6595	0.5501	0.5849	0.7361
3/20	0.4817	0.6544	0.7644	0.6644	0.6138	0.5353	0.6569	0.7639
4/20	0.2966	0.7257	0.8700	0.6980	0.6380	0.5239	0.6220	0.7847
5/20	0.2502	0.8293	0.8842	0.6376	0.6012	0.5402	0.6335	0.7986
6/20	0.1506	0.9668	0.9468	0.6309	0.5980	0.5357	0.6265	0.7014
7/20	0.0902	1.1368	0.9748	0.6711	0.5504	0.5591	0.6947	0.7639
8/20	0.0242	1.3607	0.9949	0.6443	0.5450	0.4977	0.7214	0.8333
9/20	0.0145	1.6527	0.9977	0.6711	0.5091	0.4933	0.7233	0.7778
10/20	0.0135	1.6254	0.9948	0.6577	0.4900	0.4745	0.7511	0.7986
11/20	0.056	1.6408	0.9999	0.6577	0.5177	0.5549	0.7341	0.7083
12/20	0.0108	1.7413	0.9951	0.6711	0.5222	0.6099	0.7202	0.7222
13/20	0.0174	1.7318	0.9929	0.7047	0.4453	0.6544	0.7829	0.6875
14/20	0.051	1.9297	0.9982	0.6510	0.4647	0.6032	0.7398	0.7153
15/20	0.082	1.6481	0.9982	0.6913	0.4163	0.6330	0.8114	0.7431
16/20	0.0794	1.5807	0.9674	0.6577	0.4357	0.5628	0.7900	0.6875
17/20	0.0436	1.4866	0.9790	0.6443	0.4507	0.6224	0.7658	0.7708
18/20	0.0741	1.6727	0.9740	0.6779	0.4288	0.5370	0.7859	0.7014
19/20	000538	1.4729	0.9842	0.6711	0.4204	0.5851	0.7895	0.7431
20/20	0.0144	2.0666	1.0000	0.6779	0.3888	0.6076	0.8404	0.7708

5.5   METRICS EVALUATION

The metrics such as accuracy, loss, validation loss and validation accuracy were estimated for identifying the COVID-19 patient's images for classifying the images as normal and abnormal on input CAT images.

1. **Accuracy:** It is defined as the number of correctly classified images from a total number of input CT images which makes greater enhancement in predicting CNN model performance in disease diagnosis. The formula used to measure accuracy is shown in Eqn. (1):

$$Accuracy = \frac{TP + TN}{TP + TN + FP + FN} \tag{1}$$

2. **Loss:** The loss functions are supportive to train CNN especially VGG-16 model and 3 layer CNN model. Based on actual input and predicted input images, the losses will be estimated (i.e.), distinguish among output image and target images from CAT chest images. We introduced loss function (L_{segm}), which is the integration of weighted loss (L_w) IoU and also weighted binary cross-entropy L_w BCE loss for every segmentation of images. The formula for loss function as shown in Eqn. (2):

$$L_{segm} = L_w IoU + \lambda L_w BCE \tag{2}$$

where; λ represents the weight of an image.

This L_{segm} afforded efficient image-level segmentation and pixel-level segmentation for dividing images very correctly.

3. **Validation Loss:** The loss function is validated by distinguishing among actual images and predicted images for segmenting images as normal images from COVID-19 disease patient images exactly.

4. **Validation Accuracy:** The validation accuracy is predicted by contrasting the index of the highest-scoring class in y label prediction vector and the index of the actual class in the y label true vector. It returns 0 or 1. The accuracy is predicted by correctly classified images among all input lungs CAT images.

5.6   RESULTS AND DISCUSSION

1. **Data Acquisition:** The CAT images have been gathered from the website https://medicalsegmentation.com/covid19/to find the COVID patients CT images that are helpful in medicinal health-care domains. The sample CT input images to identify the normal lung image and infected lung image are shown in Figure 5.3.

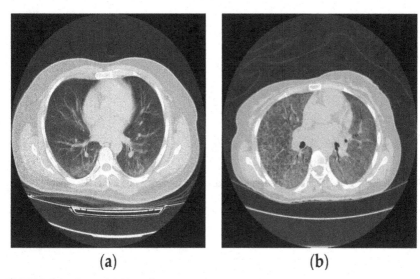

(a)   **(b)**

FIGURE 5.3   Input CAT scan images.

2. **Data Pre-Processing**: This technique performs the extraction of relevant features from lung images to identify the infected region in the lungs on CT images.
3. **Classification Using VGG-16 Model:** The images are classified as segmented images and raw images using VGG-16 DL model to estimate the training loss, validation loss, training accuracy and validation accuracy. The graph for representing model loss and accuracy on both segmented and raw images are shown in Figure 5.4.
4. **Classification of Images with Three Layer CNN Model:** The images are classified as segmented images and raw images using 3-layer CNN DL model to estimate the training loss, validation

loss, training accuracy and validation accuracy. The graph for representing model loss and accuracy on both segmented and raw images are depicted in Figure 5.5.

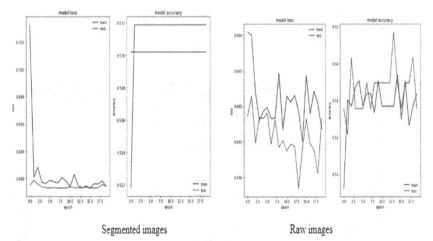

Segmented images   Raw images

FIGURE 5.4   Classification on segmented and raw images using VGG-16 model.

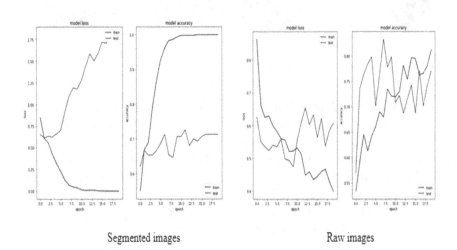

Segmented images   Raw images

FIGURE 5.5   Classification on segmented and raw images using three-layer CNN model.

5.6.1 ASSESSMENT

5.6.1.1 EFFECT OF IMAGE SEGMENTATION

By looking at the performances of the two models on both segmented images and raw images, the image segmentation allows smoother learning for both models reducing variance. And for the 3-layer CNN model, it has reached over-fitting quicker with the segmented images. However, the average validation accuracy for the segmented images is higher than the raw images. This is due to the reduction in image features during image segmentation. Additionally, a straight horizontal line for the validation accuracy of the VGG-16 model on the segmented images indicates that the model predicts the same classes for all iterations. This may be due to a lack of training or incorrect settings for weight initialization or optimization.

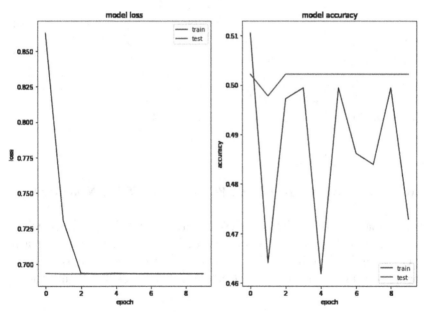

FIGURE 5.6 Loss vs. accuracy using VGG 16 model.

5.6.2 ESTIMATING LOSS AND ACCURACY USING VGG-16 MODEL

To estimate the overall performance of the used model, we are evaluating training loss along with accuracy measure and further go for the validation process to test the results. The training loss, accuracy, validation loss and accuracy for both models are depicted in Figure 5.6.

5.6.3   ESTIMATING LOSS AND ACCURACY USING CNN MODEL (FIGURE 5.7)

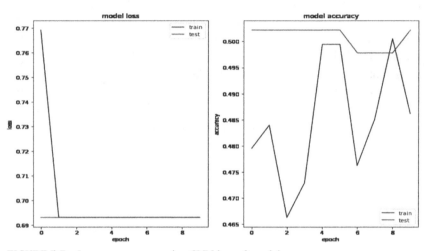

FIGURE 5.7   Loss vs. accuracy using CNN layer 2 model.

5.6.4   COMPARISON BETWEEN VGG AND 2 LAYER CNN MODELS

The two-layer CNN model on both segmented and raw images has reached over-fitting during training. However, the VGG-16 model did not reach over-fitting during training. The comparison for metrics such as loss and accuracy using VGG and layer-2 CNN model is depicted in Figure 5.8.

5.6.5   COMPARISON BETWEEN DEEP AND SIMPLE CNN MODELS

The 3-layer CNN model on both types of images has reached over-fitting during training. However, the VGG-16 model did not reach over-fitting during training. This is likely due to gradient vanishing caused by its deeper architecture. Certainly 20 epochs are not enough for the gradients to impact the first convolutions of the model. Hence, additional epochs and different hyper-parameters are given to the VGG-16 model to enhance its learning.

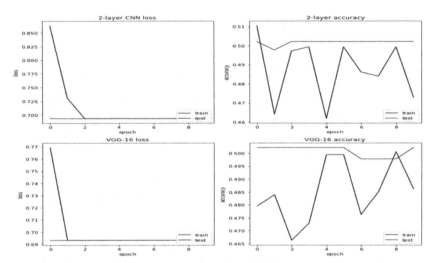

FIGURE 5.8 Comparison among CNN layer 2 and VGG model.

5.6.6 COMPARISON USING HYPERPARAMETER TUNING

The training accuracy and validation accuracy is evaluated for the model VGG-16 with 200 epochs, then tuning the parameters for attaining optimal values, and finally based on learning rate shown in Table 5.8 and the graph plot depicted in Figures 5.9 and 5.10.

TABLE 5.8 Comparison for Hyperparameter Tuning

Model	Tr_ACC	Val_ACC
VGG-16 using 200 epochs	56.31	54.86
VGG-16 with optimizer tuning	55.82	52.78
VGG-16 with reduced learning rate	52.85	54.17
Pre-trained VGG-16	93.47	69.44

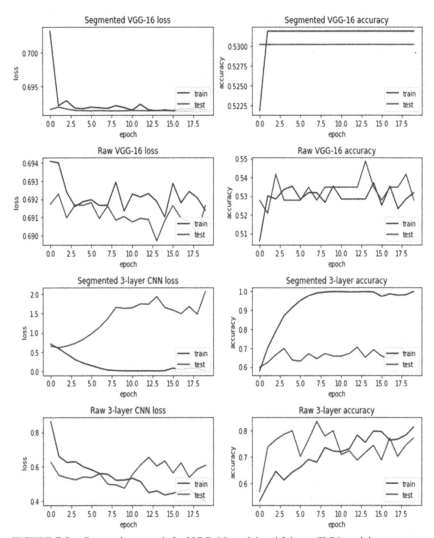

FIGURE 5.9 Comparison graph for VGG-16 model and 3 layer CNN model.

5.6.7 *OVERALL MODEL EVALUATION*

The images are classified as segmented and raw images using 2-layer CNN, DL-based VGG-16 models to categorize the images into COVID (positive) and non-COVID (negative). The overall evaluated models are illustrated in Table 5.9.

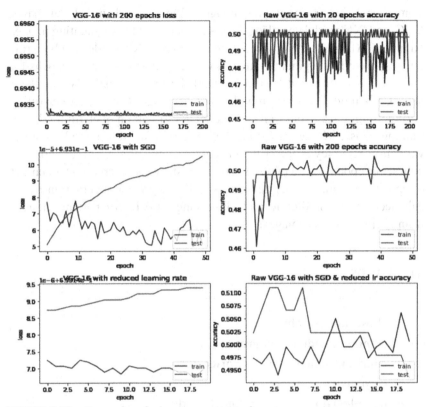

FIGURE 5.10   Comparison for hyperparameter tuning.

TABLE 5.9   Overall Model Evaluation

Models		Training Accuracy	Validation Accuracy
VGG-16	Segmented images	54.37	53.02
	Raw images	53.72	52.78
CNN 2 layer model	Segmented images	99.81	71.81
	Raw images	77.80	63.89

5.7   CONCLUSION

Coronavirus disease prediction and segmentation of COVID-19 disease identification images from other disease images had found from

Computer-Aided Tomography images. We applied novel DL-based VGG-16 CNN model and also 3 layers CNN model for segmenting images on input CAT images. Also, we introduced these CNN models on both raw images as well as segmented images via training stages for 20 epochs for validating accuracy and loss to segment the images easily. While, the 3 layer CNN model applied on both kinds of images has attain over-fitting during training stage but CGG-16 model doesn't attain over-fitting during training phase. The accuracy reaches as 100% in 3-layer model but the validation accuracy is lesser than accuracy as 67.79%. Consequently, the number of epochs is not enough in training stages, so increase in a number of epochs by tuning hyperparameters achieves greater accuracy in future enhancement in predicting and classifying COVID-19 patients images from normal on CAT images.

KEYWORDS

- chest radiography
- computer-aided tomography
- convolutional neural networks
- coronavirus disease
- deep-learning classification algorithm
- ultrasound

REFERENCES

1. Afshin, S., Marjane, K., Roohallah, A., Navid, G., Mahboobeh, J., Parisa, M., Ali, K., et al., (2020). *Automated Detection and Forecasting of COVID-19 Using Deep Learning Techniques: A Review* (pp. 1–20). arXiv:2007.10785v3 [cs.LG].
2. Ahmed, M., Congcong, W., Meng, Z., Mohib, U., Rabia, N., Hao, W., Marius, P., & Faouzi, A. C. (2020). Semi-supervised network for detection of COVID-19 in chest CT scans. *IEEE Access*, 1–14.
3. Pandit, M. K., Banday, S. A., Naaz, R., & Chishti, M. A., (2020. Automatic detection of COVID-19 from chest radiographs using deep learning. *Radiography, 27*(2), 483–489. ISSN 1078-8174, 2020.

4. Amyar, A., Modzelewski, R., Li, H., & Ruan, S., (2020). Multi-task deep learning-based CT imaging analysis for COVID-19 pneumonia: Classification and segmentation. *Computers in Biology and Medicine, 126,* 104037.

5. Michael, J. H., Subrata, C., Manoranjan, P., Anwaar, U., Biswajeet, P., Manasm S., & Nagesh, S., (2020). COVID-19 detection through transfer learning using multimodal imaging Data. *IEEE Access,* 1–17.

6. Mohammad, R., Abolfazl, A., & Seyed, M. S., (2020). A fully automated deep learning-based network for detecting COVID-19 from a new and large lung CT scan dataset, *Biomedical Signal Processing and Control,* 1–20.

7. Parisa, G., Ahmad, S., & Majid, V. (2020). Automated detection of COVID-19 using ensemble of transfer learning with deep convolutional neural network based on CT scans. *International Journal of Computer Assisted Radiology and Surgery,* 1–9.

8. Pedro, S., Eduardo, L., Guilherme, S., Gladston, M., Rodrigo, S., Diego, L., & David, M., (2020). COVID-19 detection in CT images with deep learning: A voting-based scheme and cross-datasets analysis. *Informatics in Medicine, 20,* 100427.

9. Deng-Ping, F., Tao, Z., Ge-Peng, J., Yi, Z., Geng, C., Huazhu, F., Jianbing, S., & Ling, S., (2020). *Inf-Net: Automatic COVID-19 Lung Infection Segmentation from CT Images* (pp. 1–12). arXiv:2004.14133v4 [eess.IV].

10. Maghdid, H. S., Asaad, A., Ghafoor, K., Sadiq, A., & Khan, M., (2020). *Diagnosing COVID-19 Pneumonia from X-ray and CT Images Using Deep Learning and Transfer Learning Algorithms. ArXiv, abs/2004.00038.*

11. Arnab, K. M., Sujit, K. D., Pinki, R., & Sivaji, B., (2020). Identifying COVID-19 from chest CT images: A deep convolutional neural networks based approach. *Journal of Healthcare Engineering, 2020,* 1–7. Article ID: 8843664.

12. Jamshidi, M. B., Lalbakhsh, A., Talla, J., Peroutka, Z., Hadjilooei, F., Lalbakhsh, P., & Mohyuddin, W., (2020). Artificial intelligence and COVID-19: Deep learning approaches for diagnosis and treatment. *IEEE Access,* 1–16.

13. Sadman, S., Tahrat, T., Mostafa, M. F., Zubair Md, F., & Mohsen, G., (2020). DL-CRC: Deep learning-based chest radiograph classification for COVID-19 detection: A novel approach. *IEEE Access,* 1–14.

14. Ilker, O., Boran, S., Musa, S. M., Mubarak, T. M., & Dilber, U. O., (2020). Review on diagnosis of COVID-19 from chest CT images using artificial intelligence. *Computational and Mathematical Methods in Medicine* (pp. 1–10). ISSN: 1748-6718, Article ID: 9756518.

15. Sweta, B., Praveen, K. R. M., Quoc-Viet, P., Thippa, R. G., Siva, R. K. S., Chiranji, L. C., Mamoun, A., & Md Jalil, P., (2021). Deep learning and medical image processing for coronavirus (COVID-19) pandemic: A survey. *Sustainable Cities and Society, 65,* 102589.

16. Jianpeng, Z., Yutong, X., Yi, L., Chunhua, S., & Yong, X., (2020). Viral pneumonia screening on chest X-ray images using confidence aware anomaly detection. *Accepted to IEEE Transactions on Medical Imaging.* arXiv:2003.12338v4 [eess.IV].

17. Adil, K., Khadidos, A. O., Kannan, S., Natarajan, Y., Mohanty, S. N., & Tsaramirsis, G., (2020). Analysis of COVID-19 infections on a CT image using deep sense model. *Frontiers in Public Health, 8,* 751. ISSN: 2296-2565.

18. Deepika, T. R., Keerthana, K., Ramya, T. S., & Kamalesh, S., (2020). Pneumonia detection using chest X-ray with deep learning. *International Research Journal of Engineering and Technology (IRJET), 7*(04). e-ISSN: 2395–0056 | ISSN: 2395-0072.

19. Harsh, P., Guptaa, P. K., Mohammad, K. S., & Morales-Menendez, R., Prakhar, B., & Vaishnavi, S., (2020). A deep learning and grad-CAM based color visualization approach for fast detection of COVID-19 cases using chest X-ray and CT-Scan images. *Chaos, Solitons & Fractals, 140*, 110190.

20. Yifan, P., Yu-Xing, T., Sungwon, L., Yingying, Z., Ronald, M. S., & Zhiyong, L., (2015). COVID-19-CT-CXR: A freely accessible and weakly labeled chest X-ray and CT image collection on COVID-19 from biomedical literature. *Journal Of Latex Class Files, 14*(8), AUGUST.

21. Tao, A., Zhenlu, Y., Hongyan, H., Chenao, Z., Chong, C., Wenzhi, L., Qian, T., Ziyong, S., & Liming, X., (2020). Correlation of chest CT and RT-PCR testing for coronavirus disease 2019 (COVID-19) in China: A report of 1014 cases. *Radiology, 296*(2).

22. Kai-Cai, L., Ping, X., Wei-Fu, L., Xiao-Hui, Q., Jin-Long, Y., Jin-Feng, G., & Wei, W., (2020). CT manifestations of coronavirus disease-2019: A retrospective analysis of 73 cases by disease severity, *European Journal of Radiology, 126*, 108941.

23. Frid-Adar, M., Ben-Cohen, A., Rula, A., & Hayit, G., (2018). Improving the segmentation of anatomical structures in chest radiographs using U-net with an ImageNet pre-trained encoder. *Third International Workshop*. RAMBO, arXiv:1810.02113v1 [cs.CV].

24. Rajinikanth, V., Nilanjan, D., Alex, N. J. R., Aboul, E. H., Santosh, K. C., & Madhava, R. N., (2020). Harmony-search and Otsu based system for coronavirus disease (COVID-19) detection using Lung CT scan images. *Applied Sciences*.

CHAPTER 6

Deep Learning Methods for Diabetic Retinopathy Detection

TAHIR JAVED,[1] SHEEMA PARWAZ,[2] and JANIBUL BASHIR[1]

[1]Department of Information Technology,
National Institute of Technology, Srinagar, Jammu and Kashmir, India,
E-mail: janibbashir@nitsri.ac.in (J. Bashir)

[2]Department of Computer Science Engineering, SVIET, Chandigarh,
India

ABSTRACT

Diabetes mellitus (DM) is a metabolic condition that arises because of the elevated level of blood sugar in the bloodstream. Over time, diabetes induces eye deficiency, also known as diabetic retinopathy (DR), which causes severe vision loss. This is often one of the main causes of preventable visual deficiency around the world. It has been demonstrated to influence the eyes of more than 50% of all diabetic patients to a few degrees. Beginning from obscured vision, the impacts of DR can expand to lasting visual deficiency; and in the majority of the cases, victims fail to show any early indications. The main symptoms occur in the form of augmented blood vessels, microaneurysms (MA), hemorrhages, fluid drip, and exudates in the retinal region. The conventional detection process of DR involves a trained clinician manually examining the retinal images of a patient taken by a fundus camera and looks for the nearness of vascular variations from the norm inside them. The process, by portrayal, could be a time-consuming and error-prone method that will

Application of Deep Learning Methods in Healthcare and Medical Science.
Rohit Tanwar, PhD, Prashant Kumar, PhD, Malay Kumar, PhD, & Neha Nandal, PhD (Editors)
© 2023 Apple Academic Press, Inc. Co-published with CRC Press (Taylor & Francis)

indeed in a few cases delay convenient treatment of the victim. The recent advancement in the field of machine learning has resulted in an amalgam of machine vision with deep learning (DL) techniques to perfectly train a model for correctly identifying patients with DR. In this chapter, we will discuss the different deep neural network-based frameworks to automate the DR detection process. The idea is to discuss the different artificial information models to classify the primary antecedents. We will describe the mechanisms to test the DL models using the CPU-trained neural networks (NNs). In addition, we will discuss the different mechanisms to determine the correct verdict of thresholds for determining the right class of photos of DR seriousness.

6.1   INTRODUCTION

Around 420 million individuals around the world have been analyzed with diabetes mellitus (DM). The study of this disease has shown that the same has multiplied to many folds in the last 30 years [1], and it is anticipated to extend to a larger scale. Specifically, in Asia [2], of those with diabetes, roughly one-third are anticipated to be recognized with diabetic retinopathy (DR), a persistent eye malady that can advance to irreversible vision misfortune [3]. Early discovery of this infection depends on the talented pursuers but is both work and time-devouring, and this early detainment is basic for the great forecast. In this way, it is being challenging in regions that customarily need to get to talented or skilled clinical offices. Besides, the manual strategy of DR screening advances far-reaching irregularity and mistakes among pursuers. As a result of the expanding figure within the predominance of both diabetes as well as its related retinal complications all through the world, manual strategies of conclusion are incapable of keeping pace with requests for screening administrations [4].

Computer-aided determination of DR has been investigated within the past to die down the incumbrance on ophthalmologists and relieve demonstrative irregularities between manual pursuers [5]. Computerized strategies to distinguish microaneurysms (MA) and dependably review fundoscopic pictures of DR patients have been dynamic ranges of inquiring about in computer vision of late [6]. The appearance of diverse sorts of lesions on a retina picture helps within the location of DR. These lesions include the following:

- Microaneurysms (MA) is the exceptionally most punctual clue regarding the presence of DR. It shows up as little ruddy circular dabs on the retina mainly because of frailty of the vessel's dividers. The dividers have sharp edges and are estimated to be below 125 µm.
- Hemorrhages (HM) see like bigger spots on the retina, which have a measure more noteworthy than 125 µm with an unpredictable edge. There are two different types of HM. One type is a shallow HM, called flame, and a more profound HM is called blot.
- Hard exudates: The bright-yellow spots that appear on the retina of the eye are called Hard exudates. These are caused by spillage of plasma. Hard exudates are present within the retina's external layers and they have sharp edges.
- Soft exudates show up as white spots on the retina, having oval or round shape. These are caused by the swelling of the nerve fiber. They are also called cotton wool.

The presence of massive retinal data and the availability of colossal computing power, using the graphics processing units (GPUs), have kindled the interest in deep-learning algorithms in recent years. This artificial mechanism has shown excellent results in a variety of computer vision tasks and has won by a large margin as compared to conventional methods. On similar lines, multiple deep learning (DL) methods have been created for different tasks involving the analysis of retinal fundus images in order to create automated computer-aided diagnostic systems for DR. This chapter discusses the variety of the DL models that have been developed over the time to automate the DR detection process. First, in Section 6.2, we will discuss the process of automatic DR detection process. In the same section, we will discuss the traditional and artificial DR classification methods. In Section 6.3, we will discuss the standard benchmarks that have been widely used for DR detection research. The performance metrics and loss function will be discussed in Section 6.4. Section 6.5 discusses the various DL models that have been employed in the process of DR detection. In Section 6.6, we will discuss the various prior works that have been carried in the field of DR detection using DL techniques, and finally, we will conclude in Section 6.7.

6.2   AUTOMATIC DR DETECTION PROCESS

This section presents the general overview of the different stages of DR, DR grading mechanism, detection tasks for DR and general framework to make DR detection approaches comprehensive. There is still an open and wide field of inquiry for programmed computer helped arrangements relating to DR characterization [7]. The fast retinal assessments are planned to be executed using these programmed image-based DR discovery frameworks. This will help to figure out the DR complications resulting in early diagnosis of the DR.

6.2.1   DR STAGES

Based on the seriousness of DR, it has been categorized into two different categories – non-proliferative (NPDR) and proliferative (PDR) [8, 9]. NPDR, the early stage of the two, initiates the wreckage in the retina, due to the severe damage caused to the retinal blood vessels. The damaged retina is prevalent in diabetic people. The leakage of the fluid and the blood from the damaged vessels causes the swelling of the retina. Over time, this swelling, or edema, condense the retina, resulting in vision distortion. Based on the clinical traits of this stage, at least one microaneurysm or hemorrhage with or without hard exudates is discovered [11]. Proliferative DR – and advanced sort of DR results in the generation of new blood vessels. At this point, the pathological vascular proliferation of the retina occurs, which leads to the vitreous cavity. The new blood vessels generated being very fragile, bleed into the vitreous cavity thereby resulting in vitreous hemorrhage and severe loss of vision.

6.2.2   DR GRADING

One of the critical tasks in the DR screening procedures to detect the retinal disorders is Grading. In order to identify and classify the anatomy of the DR, the evaluation and scanning of the eye is performed through ophthalmoscopy. Dilated pupils are prerequisite for this stage [12]. It thus needs a high-quality staff and a suitably sized computer screen. This is an intensive process.

Graders, such as optometrists or professionally qualified technician therapy, restore possible blindness problems, including diabetic eye disorder and macular degeneration associated with age (AMD) [12]. To perform the treatment and recovery, graders usually acquire non-mydriatic fundus images. Although if due to some media opacity the illustration is unclear, mydriatic drops are used to extend the pupil with a view to improving image clarity. All graders undergo specialized instruction based on a scanning process to ensure that the fundus images are scored consistently. These graders should spend time learning how to recognize and validate whether the cases have pathological abnormalities, as well as how to distinguish between the various levels of pathology observed. They shall make referral decisions or return for recall according to the agreed-upon timetable.

6.2.3   DR DETECTION PROCESS

The DR detection task is divided into two sub-tasks depending on the detection level: lesion-level-based and image-level-based. Every abrasion is identified, and their positions are determined in lesion-level-based detection. When deciding the magnitude of DR, the number and the location of the lesions is critical. They are crucial for determining the extent of DR severity [13]. This type of detection is further divided into two sub-phases: lesion segmentation (or detection), and lesion classification.

In fundus photographs, the distinction between the lesions like MA, hemorrhages, hard exudates, and soft exudates is done from the first place. This determines the exact location of the lesion. Retinal background pictures include similar-looking features like red dots and vessels, and so it is an overwhelming activity. The global and local contexts are used more often in this role to perform correct localization and segmentation. The identification process defines possible areas of concern. Image-based identification, on the other hand, typically focuses on image-level evaluation. This method of diagnosis is more interesting from the standpoint of scanning because it just looks for symptoms of DR [13]. This image screening task differentiates the given fundus images from natural to the ones with DR signs. This is one of the first fields of medical diagnosis where DL has contributed greatly [14].

The general identification, segmentation, and classification system include the following steps:

- Pre-processing;
- Feature extraction/selection;
- Choice of the appropriate system of classification; and
- Result evaluation.

Depending on the above steps, classification systems for detecting DR have been categorized into the following two types according to the supervised learning procedure:

- Traditional (Hand-Featured) models of pattern recognition; and
- Deep learning models of pattern recognition.

6.2.3.1 THE TRADITIONAL MODEL FOR PATTERN RECOGNITION

Since the 1950s, the traditional model for pattern recognition was based on a fixed design of a set of important features, that were manually engineered or derived from a fixed kernel (Figure 6.1). Such kernels admitted the extraction of texture, statistic, position, geometry features. These features were posteriorly combined with a simple classifier. There is a great diversity of methods for feature extraction, such as the local binary pattern, histogram of oriented gradients, gray level co-occurrence matrix, Gabor filters, among others. It was common practice to use K-nearest neighbors (KNN), linear discriminant analysis, support vector machines (SVMs) or decision trees (DTs) derived algorithms like random forest (RF) or gradient boosting, in the classification phase.

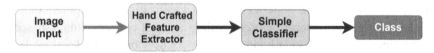

FIGURE 6.1 Traditional pattern recognition scheme.

Such a pattern recognition pipeline required a lot of labor time for designing the appropriate filters for every particular application.

6.2.3.2 DEEP LEARNING FOR PATTERN RECOGNITION

DL for pattern recognition illustrates a change of the design paradigm. A fully trainable model is designed in lieu of hand-crafting features. This fully trained model is the combination of a trainable feature extractor and a trainable classifier (Figure 6.2), creating an end-to-end automatic learner. By stacking a large number of multilayer-perceptron layers, forming deep neural networks (NNs), a concrete model is obtained. Thereby the results of ordinary NNs are profuse. But the capacity of these networks, however, are contrived by the limited processing power of computational devices. However, the training of large NNs became feasible and easy only after the introduction of high-end and computationally excellent devices like GPUs, TPUs, and frameworks like distributed systems. This has resulted in the substantial improvement in the development of Deep NNs. The precision and the usability of these models is outstanding, due to which they are used practically for a variety of problems in recent times.

The computational paradigm has been shifted from traditional CPUs to GPUs and TPU. This has ensured the facilitation of faster training of large NNs and enabled the training of CNNs possible, as well. In this way, it has paved a way for a completely different approach in image-related tasks of Machine learning. This shift along with its advancements, has resulted in the improvisation of Convolution Neural Networks (CNNs) in which, contrary to traditional image processing, the features are extracted automatically.

FIGURE 6.2   Deep learning-based pattern recognition scheme.

As a result, the CNN models have stood out to be state-of-the-art techniques for image related problems and have performed extraordinarily well in image related tasks.

6.3 DATASET

This section will present some of the standard benchmarks or datasets that have been created and developed for the automatic DR detection process. Let us briefly discuss some of these benchmarks here.

6.3.1 DIABETIC RETINOPATHY DATASET FROM KAGGLE [15]

This dataset consists of nearly 88 K images that were taken using digital fundus camera. Their resolutions vary between 433 × 289 pixels and 5,184 × 3,456 pixels. This dataset was made available to EyePACs clinics. Each image in the dataset was graded by a human reader and has been classified into four different categories:

- **Category 0:** Absence of diabetic retinopathy.
- **Category 1:** Mild NPDR.
- **Category 2:** Moderate NPDR.
- **Category 3:** Severe NPDR.
- **Category 4:** PDR.

6.3.2 DATASET FROM MESSIDOR PROJECT [16]

The Messidor was a project funded by the French government with an aim to access the different parting techniques in the domain of retinal images. The researchers under this project have developed a database containing 1,200 color fundus images of the eye with and without pupil dilation. The images were graded for the DR detection by the medical experts and have been classified into four different grades (based number of hemorrhages ($\mathcal{H}$), number of microaneurysms (μ), neovascularization ($\mathbb{N}$), and no neovascularization ($\mathbb{N}$)).

- **Grade 0 (Normal):** *($\mathcal{H}$ = 0) && (μ = 0)*
- **Grade 1:** *($\mathcal{H}$ = 0) && (0 < μ <= 5)*
- **Grade 2:** *((0 < $\mathcal{H}$ < 5) || (5 < μ < 15)) &&* $\mathbb{N}$
- **Grade 3:** *($\mathcal{H}$ >=5) || (μ > = 15) ||* N

6.3.3 DIARETDB1 DATASET [17]

The DIARETDBI database was exclusively developed for benchmarking the process of detecting DR. The database contains a total of 89 color fundus images that have been taken using the fundus camera at a 50° angle with varying settings of imaging. The images were marked by four independent medical experts. The experts mainly marked the areas of MA and hemorrhages. Based on their marks, 5 of the images were considered as normal with no sign of DR, where the rest of the images were graded in the category of mild non-proliferative DR.

6.3.4 *E-OPTHALA DATASET [18]*

The researcher associated with an ANR-TECSAN-TELEOPHTA project developed an e-opthala database containing the color fundus images. These images were examined by two independent ophthalmologists. Based on the examination, two new datasets were created: *e_optha_EX* (containing images with exudates), and *e_optha_MA* (containing images with MA. The *e_opthala_EX* has 47 exudate images and *e_optha_MA* contains 148 MA images. The entire dataset contains nearly 100 K images with resolution varying from 1,440 × 960 to 2,544 × 1,696 pixels.

Here we have briefly discussed some of the important benchmarks for DR research. However, there are many other public datasets available that are extensively used in the research of detecting DR. Some of these include: RIM-ONE dataset [19], ARIA dataset [20], SEED dataset [21], AREDS dataset [22], ORIGA dataset [23], and many more.

6.4 PERFORMANCE METRICS AND LOSS FUNCTION

6.4.1 *PERFORMANCE METRICS*

This section describes the different efficiency (or performance) metrics used to evaluate the DR detection algorithms. Some of the most frequently used metrics in this domain include F1-score, mean recall [24], mean precision [24], overall accuracy. In addition, we will also describe the notion of a loss function that characterizes a model learning behavior. Let us briefly discuss these metrics here.

1. **Mean Precision:** The potential of a classifier to not mark a negative sample as positive is measured by mean precision. It is mathematically given as:

$$Precision(for\ particular\ class) = \frac{True\ Positives\ (for\ that\ class)}{True\ Positivies\ (for\ that\ class) + False\ Positives\ (for\ that\ class)}$$

   The mean precision is determined by averaging this precision score over all the classes.

2. **Mean Recall:** The potential of a classifier to find all the positive samples in a class is measured by mean recall value. It is mathematically defined as follows:

$$Recall\left(for\ particular\ class\right) = \frac{True\ positives\left(for\ that\ class\right)}{True\ Positivies\left(for\ that\ class\right) + False\ Positives\left(for\ that\ class\right)}$$

The mean recall is determined by averaging this precision score over all the classes.

3. **Mean F1-Score:** The F1-score takes into account both precision and recall. It is defined mathematically as follows:

$$F1\ score\ \left(for\ particular\ class\right) = \frac{2\ *\ Recall\ \left(for\ that\ class\right)\ *\ Precision\ \left(for\ that\ class\right)}{Recall\ \left(for\ that\ class\right) + Precision\ \left(for\ that\ class\right)}$$

The mean F1-score is determined by averaging this precision score over all the classes.

4. **Overall Accuracy:** It is an easy way to evaluate a model's efficiency. It is the proportion of correctly categorized items in the test set to the total number of items in the test sample.

6.4.2   LOSS FUNCTION

Loss function is a means of determining how efficiently a model is performing on given data. Networks learn using the loss function. If the model prediction is far away from the ground truth, in that case loss function will depict a large value, on the other hand, it will be less. So, the whole learning of the network can be concluded as minimizing the loss incurred. For most classification tasks cross-entropy loss is used specifically, otherwise one can typically go along with any loss function.

Cross entropy can be understood as follows:

Entropy as per Claude Shannon is the minimum encoding size to efficiently send messages without losing any information. The entropy of a discrete probability distribution is as follows:

$$Entropy\ \sum P(i) \times Log(P(i))$$

The probability *P* for each *i* is assumed to be known. The term *i* is a discrete event.

For continuous variables, the Entropy can be defined using the following integral:

$$\text{Entropy} \int P(x) \times \text{Log}(P(x))$$

here; x is a continuous variable; and P(x) is the probability density function. In both cases, the actual expectation of the negative log probability is calculated which is the theoretical minimum encoding size of the information from the event x. So, entropy can be written as follows:

$$\text{Entropy Ex} \sim P \times [-\text{Log}(P(x))]$$

Now, if actual distribution P is not known, another distribution Q can be used as an estimate to calculate entropy which is called Estimated Entropy.

$$\text{Estimated Entropy Ex} \sim P \times [-\text{Log}(Q(x))]$$

If Q is close to P the above value will be a close estimate of the smallest encoding size. But if real distribution P is known and also the estimated distribution Q, Cross-Entropy can be calculated, which gives the difference between the two distributions or gives a quantitative description of how close the two distributions are. Cross-Entropy is calculated as:

$$\text{Cross Entropy Ex} \sim P \times [-\text{Log}(Q(x))]$$

This is the proper way in which cross-entropy is used as a loss function for training the NNs. Where in, the ground truth label for the inputs is the actual distribution P and the output produced by the model is the estimated distribution Q and the deviation between these two distributions is quantified by Cross-Entropy, which is the loss incurred by the network. In this way we get a probabilistic interpretation of the output produced by the network, which is the main advantage of using Cross-Entropy as the loss function. It tries to push the correct class probability all the way to 1 and others to 0.

6.5 DEEP LEARNING ARCHITECTURES

A number of DL-based architectures have been implemented over time. Among these architectures, Convolution Neural Networks (CNNs) have been most widely used in the process or DR detection. Apart from CNNs, there are many other DL architectures such as recurrent neural networks

(RNNs), autoencoders (AEs), and deep belief networks, that have been used in the DR detection. Let us briefly discuss these architectures here.

6.5.1 CONVOLUTIONAL NEURAL NETWORKS (CNNS)

CNNs are specialized in exploiting the natural local correlations present in images. They are able to create local abstractions that are combined layer-by-layer forming more elaborated meta-abstractions, hopefully, being able to disentangle the information contained in the image, forming features that are useful for solving a particular classification/regression problem. The fundamental operator that is used to capture this correlation in the images is the convolution operator.

Broadly the CNN can be studied as a two-part unit:

- Feature extraction part; and
- Fully connected part.

The first part has a large number of convolution layers (CLs) stacked on top of each other. These CLs extract features in a way similar to that of the human visual system. The initial CLs are responsible for capturing the simple features among the pixels like lines. The subsequent CLs use those simple abstractions and extract more complex relationships.

Each layer has a convolutional operator that transforms input representation to an output representation depending on the filter used by the convolution operation. The output can pass through other transformations inside the same layer, like a batch-normalization, an activation function or reducing size functions like max-pooling or average pooling blocks. All these functions have learnable parameters that are tuned through a loss function optimization process with the objective of enhancing the classification/regression capabilities of the network. The characteristic building block of such a type of network is the convolution operator.

1. **Pooling Operator/Layer:** It is a typical operator used often in the context of CNNs. Pooling operator essentials help in reducing the number of trainable parameters. It is also called subsampling or down-sampling. It is mainly required for dimensionality reduction of each feature map [25]. It also helps in the reduction of computation work in the network. Pooling can be of the following types:
   - Max pooling;

- Average pooling;
- Sum pooling.
2. **Non-Linearity:** It refers to applying a nonlinear function at every layer of CNN in order to allow the model to learn complex mappings from inputs to outputs.

So, in essence, a CNN model learns the hierarchy of features automatically and then produces excellent classification performance; in this way, the model is trained from beginning to end. A deep model that takes inputs and generates outputs is generally referred to as an end-to-end model. Transfer learning is another mode of learning that CNNs can use. This means that a model is first taught end-to-end using a dataset from a different domain and then fine-tuned using the domain dataset.

A large amount of data is needed for a CNN model to work properly, as it must resolve overfitting issues and ensure proper convergence [26]. Large amounts of data, especially for DR detection, are not available in the medical domain. As a result, transfer learning [27] is the solution. In general, there are two transfer learning methods that are used:

- As a function extractor, a pre-trained CNN model; and
- Using data from the appropriate domain to fine-tune a pre-trained CNN model.

A completely convolutional network (FCN) is a CNN model in which CLs are created from fully connected layers and in order to reverse the effect of CLs (i.e., downsampling), deconvolution (or transposed convolution) layers are added. This is done so that the output image has the same size as the input image [28]. Segmentation is mostly done using this specific model.

6.5.2 METHODS FOCUSED ON AUTOENCODERS AND STACKED AUTOENCODERS

A NN with a single hidden layer assembled in such a way that the output is the same as the input is called an autoencoder (AE) [29]. This is used to construct a stacked-autoencoder (SAE), a complex architecture [30]. Pre-training and fine-tuning are the two stages of training for such SAE models. An SAE is trained layer by layer, but unattended, during the pre-training process. The pre-trained SAE model is now fine-tuned in the fine-tuning process, using gradient descent and backpropagation algorithms in an attended manner. An AE is the fundamental component of an SAE. AEs are classified into two types: sparse and denoising. For extracting sparse features from the raw data, Sparse autoencoders are used. The representation's sparsity can be obtained by either penalizing hidden unit biases or explicitly penalizing the performance of hidden unit activations. Maji et al. [31] used denoising autoencoders (DAEs) in DR detection in order to catch the right version of data from the corrupted input data.

6.5.3 RECURRENT NEURAL NETWORKS (RNNS)

RNNs are a category of NNs that involve iterations. It learns both the context as well as the input patterns. The output is generated by the previous iterations of the output that gets integrated with the current input. These recurrences affect the RNN. Typically, an RNN model has three sets of parameters: input to hidden weights W, hidden weights U, and hidden weights to output, with weights spread around positions/times in the input series [32].

6.5.4 DEEP BELIEF NETWORKS

DBNs are deep network architectures composed of cascading restricted Boltzmann machines (RB) [33]. For an RBM, the similarity between the input and its projection needs to be maximized which is executed by using a contrastive divergence algorithm. DBNs have probabilistic nature. To avoid any kind of degenerate solutions they use probability as a similarity metric. The DBNs are trained in an unsupervised manner using a layer-by-layer greedy learning technique. This technique uses the gradient descent and backpropagation algorithms in order to fine-tune the final architecture.

6.6 PREVIOUS WORKS

The first artificial neural networks (ANN) pertaining to this field of research investigated the capacity to sort out different patches within the retina. The patches include the regions with and without blood vessels, regions with exudates and pathologic retinas with MA. Eventually, compared to normal patches of the retina, an accuracy of about 74% was being able to detect MA [34]. Various previous experiments have been performed using high bias, low variance optical image processing techniques. These have done a good job of defining one particular function used in the diagnosis of subtle illness, such as the use of the top-hat algorithm to identify MA [35–37]. Apart from MA, there are a number of other characteristics that make it difficult to miss this condition. As seen in the recent past, there has been considerable advancement in the field of machine learning and its applicability to a number of ventures. Some of the works done in the field of Retinopathy are summed as:

GiriBabu et al. [38] presented Spatially Weighted Fuzzy c-Means Clustering for Vessel Segmentation in Fundus Images for extracting Blood Vessels. The researchers used a series of linear filters that were adaptive to vessels of various thicknesses and orientations. There is a very simple method reported in the literature: a vessel detection method and an experimental evaluation of this method has demonstrated excellent performance over global thresholding.

Bob et al. [39] described a method for Detecting Diabetes Mellitus and NPDR. They used Geometry Features, Color, and Texture for detection via this method. First, the image is captured and pre-processed. Then color features of the captured image are extracted from the pre-processed image. The image is separated into eight blocks, in order to represent the texture feature. The 2-D Gabor filter, named after Dennis Gabor, is applied on the surface for compute texture analysis. Finally, the geometry features which include distances, areas are extracted from the foreground image. For separating NPDR samples from healthy samples, the SVM classifier is used. However, this method has achieved only 80.52% accuracy so far.

Yuji et al. [40] demonstrated that the use of brightness correction on fundus images boosts the method of automated hemorrhage detection. This way indicates the importance of developing several automated models in order to find out the abnormalities in fundus images. The purpose of the chapter presented by them was to mend their automated hemorrhage detection model for the proper diagnosis of DR. They showed a new method for pre-processing as well as for false positive (FP) elimination. By using a 45-feature analysis, they eliminated the false positives. They looked at 125 fundus photographs, 35 of which had hemorrhages and 90 of which were fine, to see the efficacy of this new approach. The sensitivity and accuracy for detecting rare cases in this system were 80% and 88%, respectively.

These validated findings indicate that their new approach can significantly improve the efficiency of their hemorrhage diagnosis system.

On a dataset of 55 patients, Vujosevic et al. [41] build a binary classifier by directly forming single lesion attributes. However, the focus of this research is limited to the dataset.

For blood vessel segmentation, Wang et al. [42] used a CNN (LeNet-5 architecture) as a function extractor. Three heads were placed at various layers of the convnet in this model. They are then fed into three random trees, which are then assembled for a final forecast by the final classifier. This model achieved an accuracy and AUC on 0.97/0.94 using a standard dataset for comparing models which addressed vessel segmentation.

The authors of Ref. [43] explained how they developed a CNN for lesion-level classification. The studied attribute representation is then used to classify images at the level of detail. The study's reach is restricted, however, since the dataset only contains 200 photographs.

The research work presented in Ref. [44] contains morphological image processing methods for removing blood vessels, MA, exudate, and hemorrhage characteristics. After that, an SVM was trained on a data collection of 331 photographs. This achieved sensitivity 82% and specificity 86%.

Using 140 image dataset, Bhat et al. [45] reported sensitivity of 90% and accuracy of 90%. They remove the region of blood vessels, exudates, and texture characteristics using image processing techniques. To obtain the final result, these are fed into a small NN.

As a result, the need for a robust and automated DR screening system has long been recognized. In this chapter, we have described the different automatic DL-based approaches that can aid to automate the tedious, manual, and lengthy process of retinopathy detection. We will also consider different causal factors of DR.

6.7 CONCLUSION

Detecting DR in humans using traditional methods under the guidance of expert ophthalmologists is complex and time-consuming. The shortage of specialists and the amount of time it takes to complete the procedure will cause a delay in identifying illness and, as a result, a delay in care, which can result in vision loss in certain cases. CNNs have been shown to be highly efficient at identifying physically observable patterns in images, and the advancement of computing hardware has made their use feasible. Automated diagnosis and scanning provide a once-in-a-lifetime chance to avoid a large portion of our population's vision loss. CNNs have recently been introduced to the list of algorithms used to screen for diabetic disease by researchers. CNNs promise to benefit from raw pixels by using the massive quantities of images amassed for physician-interpreted scanning. CNNs may be able to detect a broader variety of non-diabetic diseases due to the high variance and low bias of these models. Using CNNs to detect DR in humans can result in better detection and on time therapies to mitigate the disease's effects on humans.

This chapter dealt with different methods for automated DR detection, however, the challenges are subtle, particularly in the case when the disease symptoms are not very clear, not even to human practitioners. We have seen how deep-learning especially CNNs have been remarkable in their process and capable of making sense of critical data. These systems yield remarkable results in a way that despite having only a subset of data and limitations of hardware, the CNNs models have proved to perform exceptionally well. However, if there are no bottlenecks of limitations and more verbose dataset is available, it is certain that such tasks in medical fields will be revolutionized with newer and robust DL models.

KEYWORDS

- autoencoder
- convolution neural networks
- diabetic retinopathy
- graphics processing units
- hemorrhages
- macular degeneration associated with age
- microaneurysms
- non-proliferative
- recurrent neural networks

REFERENCES

1. WHO, (2016). *Global Report on Diabetes.*
2. Chan, J. C., Malik, V., Jia, W., Kadowaki, T., Yajnik, C. S., Yoon, K. H., & Hu, F. B., (2009). Diabetes in Asia: epidemiology, risk factors, and pathophysiology. *JAMA, 301*(20), 2129–2140.
3. Congdon, N. G., Friedman, D. S., & Lietman, T., (2003). Important causes of visual impairment in the world today. *JAMA, 290*(15), 2057–2060.
4. Goh, J. K. H., Cheung, C. Y., Sim, S. S., Tan, P. C., Tan, G. S. W., & Wong, T. Y., (2016). Retinal imaging techniques for diabetic retinopathy screening. *Journal of Diabetes Science and Technology, 10*(2), 282–294.
5. Mookiah, M. R. K., Acharya, U. R., Chua, C. K., Lim, C. M., Ng, E. Y. K., & Laude, A., (2013). Computer-aided diagnosis of diabetic retinopathy: A review. *Computers in Biology and Medicine, 43*(12), 2136–2155.
6. Philip, S., Fleming, A. D., Goatman, K. A., Fonseca, S., Mcnamee, P., Scotland, G. S., Prescott, G. J., et al., (2007). The efficacy of automated "disease/no disease" grading for diabetic retinopathy in a systematic screening program. *British Journal of Ophthalmology, 91*(11), 1512–1517.
7. Mansour, R. F., (2017). Evolutionary computing enriched computer-aided diagnosis system for diabetic retinopathy: A survey. *IEEE Reviews in Biomedical Engineering, 10*, 334–349.
8. Early Treatment Diabetic Retinopathy Study Research Group, et al., (1991). Grading diabetic retinopathy from stereoscopic color fundus photographs an extension of the modified Airlie House classification: ETDRS report number 10. *Ophthalmology, 98*(5), 786–806.
9. Udyavara, R. A., Choo, M. L., Yin, K. N. E., Caroline, C., & Toshiyo, T., (2009). Computer-based detection of diabetes retinopathy stages using digital fundus images. *Proceedings of the Institution of Mechanical Engineers, Part H: Journal of Engineering in Medicine, 223*(5), 545–553.
10. *Diabetic Retinopathy.* https://www.nhs.uk/conditions/diabetic-retinopathy/stages/ (accessed on 14 May 2022).
11. Alan, D. F., Keith, A. G., Sam, P., Graeme, J. W., Gordon, J. P., Graham, S. S., Paul Mc, N., Graham, P. L., William, N. W., Peter, F. S., et al., (2010). The role of hemorrhage and exudate detection in automated grading of diabetic retinopathy. *British Journal of Ophthalmology, 94*(6), 706–711.
12. *Diabetic Retinal Screening, Grading, Monitoring, and Referral Guidance.* https://www.health.govt.nz/publication/diabetic-retinal-screening-grading-monitoring-and-referral-guidance (accessed on 14 May 2022).
13. Lama, S., Thomas, H., Jihed, C., Farida, C., & Pierre, L. J. M., (2016). Red lesion detection using dynamic shape features for diabetic retinopathy screening. *IEEE Transactions on Medical Imaging, 35*(4), 1116–1126.
14. Geert, L., Thijs, K., Babak, E. B., Arnaud, A. A. S., Francesco, C., Mohsen, G., Jeroen, A. W. M. V. D. L., Bram, V. G., & Clara, I. S., (2017). A survey on deep learning in medical image analysis. *Medical Image Analysis, 42*, 60–88.
15. Kaggle Dataset. https://www.kaggle.com/c/diabetic-retinopathy-detection/data (accessed on 14 May 2022).

16. Messidor Dataset. http://www.adcis.net/en/Download-Third-Party/Messidor.html (accessed on 14 May 2022).
17. Kälviäinen, R. V. J. P. H., & Uusitalo, H., (2007). DIARETDB1 diabetic retinopathy database and evaluation protocol. In: *Medical Image Understanding and Analysis* (Vol. 2007, p. 61).
18. Decenciere, E., Cazuguel, G., Zhang, X., Thibault, G., Klein, J. C., Meyer, F., Marcotegui, B., et al., (2013). TeleOphta: Machine learning and image processing methods for teleophthalmology. *IRBM, 34*(2), 196–203.
19. Fumero, F., Alayón, S., Sanchez, J. L., Sigut, J., & Gonzalez-Hernandez, M., (2011). RIM-ONE: An open retinal image database for optic nerve evaluation. In: *2011 24ᵗʰ International Symposium on Computer-Based Medical Systems (CBMS)* (pp. 1–6). IEEE.
20. Aria Dataset. http://www.eyecharity.com/aria_online.html (accessed on 14 May 2022).
21. Pan, C. W., Zheng, Y. F., Anuar, A. R., Chew, M., Gazzard, G., Aung, T., Cheng, C. Yet al., (2013). Prevalence of refractive errors in a multiethnic Asian population: The Singapore epidemiology of eye disease study. *Investigative Ophthalmology & Visual Science, 54*(4), 2590–2598.
22. Nih AREDS Dataset. https://www.nih.gov/news-events/news-releases/nih-adds-first-images-major-research-database (accessed on 14 May 2022).
23. Zhang, Z., Yin, F. S., Liu, J., Wong, W. K., Tan, N. M., Lee, B. H., Cheng, J., & Wong, T. Y., (2010). Origa-light: An online retinal fundus image database for glaucoma analysis and research. In: *2010 Annual International Conference of the IEEE Engineering in Medicine and Biology* (pp. 3065–3068). IEEE.
24. https://www.biostat.wisc.edu/~page/rocpr.pdf (accessed on 14 May 2022).
25. Scherer, D., Müller, A., & Behnke, S., (2010). *Evaluation of Pooling Operations in Convolutional Architectures for Object Recognition*, 92–101.10. 1007/978-3-642-15825-4_10.
26. Nima, T., Jae, Y. S., Suryakanth, R. G., Todd, H. R., Christopher, B. K., Michael, B. G., & Jianming, L., (2016). Convolutional neural networks for medical image analysis: Full training or fine-tuning? *IEEE Transactions on Medical Imaging, 35*(5), 1299–1312.
27. Dumitru, E., Yoshua, B., Aaron, C., Pierre-Antoine, M., Pascal, V., & Samy, B., (2010). Why does unsupervised pre-training help deep learning? *Journal of Machine Learning Research, 11*, 625–660.
28. Jonathan, L., Evan, S., & Trevor, D., (2015). Fully convolutional networks for semantic segmentation. In: *Proceedings of the IEEE Conference on Computer Vision and Pattern Recognition* (pp. 3431–3440).
29. Geoffrey, E. H., & Ruslan, R. S., (2006). Reducing the dimensionality of data with neural networks. *Science, 313*(5786), 504–507.
30. Cheng-Yuan, L., Wei-Chen, C., Jiun-Wei, L., & Daw-Ran, L., (2014). Autoencoder for words. *Neurocomputing, 139*, 84–96.
31. Debapriya, M., Anirban, S., Sambuddha, G., Debdoot, S., & Pabitra, M., (2015). Deep neural network and random forest hybrid architecture for learning to detect retinal vessels in fundus images. In: *Engineering in Medicine and Biology Society (EMBC), 2015 37ᵗʰ Annual International Conference of the IEEE* (pp. 3029–3032). IEEE.

32. Tom´aˇs, M., Martin, K., Luk´aˇs, B., Jan, C., & Sanjeev, K., (2010). Recurrent neural network-based language model. In: *11ᵗʰ Annual Conference of the International Speech Communication Association.*
33. Oriol, V., Alexander, T., Samy, B., & Dumitru, E., (2017). Show and tell: Lessons learned from the 2015 mscoco image captioning challenge. *IEEE Transactions on Pattern Analysis and Machine Intelligence, 39*(4), 652–663.
34. Gardner, G., Keating, D., Williamson, T., & Elliott, A., (1996). Automatic detection of diabetic retinopathy using an artificial neural network: A screening tool. *British Journal of Ophthalmology, 80*(11), 940–944.
35. Niemeijer, M., Van, G. B., Cree, M. J., Mizutani, A., Quellec, G., Sanchez, C. I., Zhang, B., Hornero, R., Lamard, M., Muramatsu, C., et al., (2010). Retinopathy online challenge: automatic detection of microaneurysms in digital color fundus photographs. *IEEE Transactions on Medical Imaging, 29*(1), 185–195.
36. Wang, S., Tang, H. L., Hu, Y., Sanei, S., Saleh, G. M., Peto, T., et al., (2017). Localizing microaneurysms in fundus images through singular spectrum analysis. *IEEE Transactions on Biomedical Engineering, 64*(5), 990–1002.
37. Mookiah, M. R. K., Acharya, U. R., Chua, C. K., Lim, C. M., Ng, E. Y. K., & Laude, A., (2013). Computer-aided diagnosis of diabetic retinopathy: A review. *Computers in Biology and Medicine, 43*(12), 2136–2155.
38. Giri, B. K., Satya, S. T., & Subbaiah, P. V., (2007). Segmentation of vessels in fundus images using spatially weighted fuzzy C-means clustering algorithm. *International Journal of Computer Science and Network Security, 7*(12), 102–109.
39. Bob, Z., Vijaya, K. B. V. K., & David, Z., (2014). Detecting diabetes mellitus and non proliferative diabetic retinopathy using tongue color, texture, and geometry features. *IEEE Transactions on Biomedical Engineering, 61*(2), 491–501.
40. Yuji, H., Toshiaki, N., Yoshinori, H., Masakatsu, K., Akira, S., Kazuhide, K., Takeshi, H., & Hiroshi, F., (2008). Improvement of automatic hemorrhages detection methods using brightness correction on fundus images. In: *Proc. of the International Society of Optics and Photonics on Medical Imaging* (Vol. 6915).
41. Vujosevic, S., Benetti, E., Massignan, F., Pilotto, E., Varano, M., Cavarzeran, F., Avogaro, A., & Midena, E., (2009). Screening for diabetic retinopathy: 1 and 3 nonmydriatic 45-degree digital fundus photographs vs 7 standard early treatment diabetic retinopathy study fields. *American Journal of Ophthalmology, 148*(1), 111–118.
42. Wang, S., Yin, Y., Cao, G., Wei, B., Zheng, Y., & Yang, G., (2015). Hierarchical retinal blood vessel segmentation based on feature and ensemble learning. *Neurocomputing, 149*, 708–717.
43. Lim, G., Lee, M. L., Hsu, W., & Wong, T. Y., (2014). Transformed representations for convolutional neural networks in diabetic retinopathy screening. In: *Workshops at the 28ᵗʰ AAAI Conference on Artificial Intelligence.*
44. Acharya, U. R., Lim, C. M., Ng, E. Y. K., Chee, C., & Tamura, T., (2009). Computer-based detection of diabetes retinopathy stages using digital fundus images. *Proceedings of the Institution of Mechanical Engineers, Part H: Journal of Engineering in Medicine, 223*(5), 545–553.
45. Yun, W. L., Acharya, U. R., Venkatesh, Y. V., Chee, C., Min, L. C., & Ng, E. Y. K., (2008). Identification of different stages of diabetic retinopathy using retinal optical images. *Information Sciences, 178*(1), 106–121.

CHAPTER 7

Study to Distinguish Covid-19 from Normal Cases Using Chest X-Ray Images with Convolution Neural Network

P. SATHISH KUMAR,[1] P. VIJAYALAKSHMI,[2] and V. RAJENDRAN[2]

[1]*Department of Electronics and Communication Engineering, Bharath Institute of Higher Education and Research, Selaiyur, Chennai-73, Tamil Nadu, India, E-mail: sathishmrl30@gmail.com*

[2]*Department of Electronics and Communication Engineering, Vels Institute of Science, Technology, and Advanced Studies (VISTAS), Chennai, Tamil Nadu, India, E-mail: sathish.se@velsuniv.ac.in (P. S. Kumar)*

ABSTRACT

An official announcement by the United Nations agency has confirmed an enduring deadly disease, namely coronavirus (COVID-19). While comparing to traditional methods such as blood tests and analysis, chest X-ray (CXR) images are the most important evidential techniques to find whether the patient is normal or COVID-19. In order to distinguish the COVID-19 patient among these images there is a classification method via convolution neural network (NN) techniques adapted. In this chapter, the experimental outcomes in which histogram equalization approach is utilized to process the input CXR images. To find accurate results deep learning (DL)-based convolution neural network (CNN) achieves accuracy as 50%, subsequently reshaping the input images into 224×224 along

Application of Deep Learning Methods in Healthcare and Medical Science.
Rohit Tanwar, PhD, Prashant Kumar, PhD, Malay Kumar, PhD, & Neha Nandal, PhD (Editors)

with batch size 32 undergoes training phase and it consists of 10 epochs to generate enhanced accuracy as 96.4% by tuning hyperparameters. Moreover, we applied optimization algorithm such as Adam optimizer, RMSprop optimizer and stochastic gradient descent (SGD) in which Adam and SGD achieves validation accuracy as 96.67% yields better performance in classifying images as normal and Coronavirus disease patients.

7.1 INTRODUCTION

COVID-19 is an extremely contagious disease imputable to cruel sensitive respiratory disorder, which is referred as Coronavirus. The disease foremost initiated from Wuhan city located in China, December 2019 ever since that disease increase worldwide influencing around 200 countries. The main impact of public health emergency of international concern is officially announcement by World Health Organization (WHO) has confirmed the enduring deadly disease of COVID-19. As of 30th October 2020, a total of 52,681,305 COVID-19 cases with death rate as 1,292,279 in more than 200 countries across the world depicted in Figure 7.1.

FIGURE 7.1 COVID-19 cases as of 30th October 2020.

Therefore, in this particular situation, the foremost obsession needs to be done also most of deadly disease affected countries started manual testing; accordingly suitable decisions can be taken based on correct situation. Some shortcomings while doing physical test:

- Spare accessibility on disease trial equipment;
- Expensive and also incompetent bloodline trial;
- Moreover, bloodline analysis acquires nearly 5–6 hours to produce blood report.

Hence, to defeat these kinds of manual testing issues, we choose DL approach mainly convolutional neural network for improved and proficient treatment such as generate patients report earlier because of disease is exceedingly transmittable.

7.2 RELATED WORK

Adedoyin et al. [1] demonstrate how big data especially AI technique helpful for medicinal diagnosis such as predicting COVID-19 disease, pneumonia, SARS, likewise. The examined techniques published advances in medicinal data investigation with an accuracy of equal to 90%. Albahri et al. [5]; and Afshin [2] offered a systematic review of artificial intelligence (AI) techniques applicable in detecting disease mainly coronavirus via mammogram (X-ray) images and computed tomography (CT) images. Furthermore, this review suggested a comprehensive methodology for estimation along with guidelines of AI techniques utilized in the entire classification responsibilities of COVID-19 medicinal picture as outlook ways.

Narinder et al. [21]; and Ahmad et al. [3] introduced how DL and machine learning models helpful in predicting COVID-19 disease patients around worldwide which is applicable for synchronized information. Boran et al. [7] scope is to utilize DL methods for achieving high accuracy detection of COVID-19 through CXR images by means of applying machine learning and DL algorithms. Ekta et al. [8] applied machine learning algorithms such as support vector machine (SVM) and polynomial regression for finding total confirmed COVID-19 cases in several states of India. Elaziz et al. [9] collected CXR images for classifying COVID-19 person and normal had analyzed by feature extraction using novel fractional multichannel exponent moments (FrMEMs) and modified fractional multichannel exponent moments. Feng et al. [10] insisted AI for efficiency enhancement in the presentation of infections in both X-ray images and also CT images. Here, the methods utilized for diagnose patients disease through four steps, namely image acquisition, segmentation, diagnosis, and finally follow-up. Irfan et al. [13] objective is to analyze COVID-19 patients using deep CNN approach namely LeNet, VGG-16, Inception, ResNet50. From this work, the author concludes that people protect themselves from somebody who has more publicity to virus or symptoms of COVID-19.

Jingxin et al. [15] integrated DL target detection as well as image classification methods to learn CT scan images of COVID-19 patients. This survey developed recurrent neural network (RNN) and 2D convolutional network for automatic detection of COVID-19 cases by getting patients CT images. Michael et al. [16] creates CNN algorithm especially VGG19 to afford over-stressed medicinal proficient a second pair of eyes through patients X-rays, CT scans, and also ultrasound images by finding metrics like precision. Mohammad et al. [17] applied AI base techniques in categorization of images, study, and enhancement of medicinal images. To achieve this goal, the researcher developed DL algorithms namely RNN, long short term memory (LSTM), generative adversarial networks (GAN), and ELM to diagnose COVID-19 patients. Mohammad et al. [18] proposed deep Convolutional networks for classifying X-ray images into pneumonia, normal, and COVID-19. The total number of images taken for input as 180 X-ray images and construct CNN model integrated with Xception and ResNet50V2. The performance estimated through accuracy prediction for all classes, especially COVID-19 disease too.

Tengku et al. [24] demonstrate how IEEE 802.11 ah net helpful in COVID-19 anticipation group (doctors, nurses, and also governments). By finding frequency, Bandwidth, and Maximum data rate for various IEEE networks mainly for detecting COVID-19 cases. Wael et al. [25] surveyed recognition of patients with alleged coronavirus infection via phone calls. Moreover, this study constructed support vector machine (SVM) and also ANN for judgment of COVID-19 patients as well as comparison of performances were analyzed. Finally, investigations were done on Arabic speech signals of recorded phone calls. From that signals, appropriate features extraction of keywords is accomplished.

Wang et al. [26] projected an entirely repeated DL approach for COVID-19 investigative and predictive investigation by regularly used to work out tomography. Xiang et al. [27] datasets collected from pulmonary hospital, Wuhan City between January 3 to February 13, 2020 applied

DL methods to recognize mild patients with probable cruel development very efficiently and accurately. In addition to that, logistic regression approaches were built to classify COVID-19 patients and normal patients. This survey concluded that DL-based approach produced better accuracy when compared to logistic regression.

Some of various literature papers regarding how COVID-19 disease is analyzed using various techniques, also describe the objective of work, what kind of input have taken, finally outcomes generated by several researchers shown in Table 7.1.

TABLE 7.1 Surveys Done by Several Researchers About COVID-19

Survey	Objective of Work	Input Taken by Several Researchers	Methodology Used	Output
Javier et al. [14]	Diagnose patients affected by coronavirus	Mammogram Pulmonary images	VGG16 based deep learning method	Finding measures such as recall, precision, and accuracy.
Muammer et al. [20]	Classifying input images as normal, COVID-19, and pneumonia	X-ray images	COVIDetectioNet model	Finding accuracy to evaluate performance in classifying images
Amit et al. [6]	Detect COVID-19 positive patients using chest X-ray images	Chest X-ray images	Deep CNN, develop GUI application too	Diagnose COVID-19 positive patients via evaluating classification accuracy
Parul et al. [22]	Predicting and analyzing COVID-19 positive cases	COVID-19 positive report cases	RNN, LSTM	Applicable in real-time information which is used in several countries
Harit et al. [11]	Finding performance in detection of COVID-19	Chest X-ray radiographs images as input	ResNet50, fully connected layer, rectified linear unit	Determining accuracy as metrics in calculating performance.

TABLE 7.1 *(Continued)*

Survey	Objective of Work	Input Taken by Several Researchers	Methodology Used	Output
Moutaz et al. [19]	Examine chest X-ray images	1,000 X-ray images	Deep CNN Prophet algorithm, autoregressive integrated moving average, long short term memory	Classify normal images from COVID-19 X-ray images
Rachna et al. [23]	Detection and analysis of COVID-19	Chest X-ray scans	Inception V3, Xception, and ResNeXt algorithm in CNN	Detecting COVID-19 patients via chest X-ray images.
Harsh et al. [12]	Analyzing X-rays for detecting COVID-19 patients	CT scan images and X-ray images	nCOVNet model	Detecting loss and accuracy in diagnosis of COVID-19
Ahmed et al. [4]	Classifying COVID-19 images carried out by feature extraction	COVID-19 X-ray of two datasets	Marine predators algorithm	Enhanced in performance and reduction of computational complexity

7.3 PROPOSED METHODOLOGY

Figure 7.2 demonstrates the workflow of our proposed work in distinguishing input chest X-ray (CXR) images as COVID-19 and normal via layers in CNN.

7.3.1 PREPARING DATA

The original images have been gathered from specified source followed by that train data, process it, and finally produce outcome to predict the model performance. Hence, we applied DL techniques to diagnose the disease namely COVID-19 classification from patients CXR images.

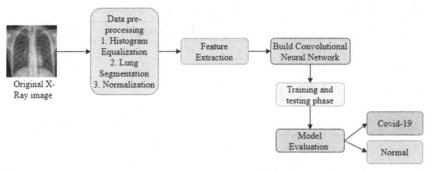

FIGURE 7.2   Workflow for proposed work.

- Chest mammogram images gathered from Kaggle website to mine mammograms (X-rays) of vigorous patient's and also have 284 trial images necessitate to balance with Coronavirus obtainable images.
- Also, the author of this chapter has considered Coronavirus patients datasets especially mammogram images.
- Nowadays, poster anterior (PA) view is the most recurrently used for viewing images; hence in this proposed work, the author of this chapter apprised the coronavirus disease patients using PA analysis mammogram scans for my investigation.

7.3.1.1   DATASETS INFORMATION

The proposed work utilized two kinds of datasets for detecting and classifying images as normal, coronavirus (COVID-19), and gathered patient's details who suffered from pneumonitis via getting chest mammogram images. The description of datasets as well as splitting phases (training and testing) is detailed in Tables 7.2 and 7.3.

TABLE 7.2   Datasets Information

SL. No.	Datasets
1.	COVID Chest X-ray (*Source:* https://github.com/ieee8023/covid-chestxray-dataset)
2.	Pneumonia chest X-ray (*Source:* https://www.kaggle.com/paultimothymooney/chest-xray-pneumonia)

TABLE 7.3   Dataset Class Allocation for Both Training and Testing Phase

Category	Training Stage	Testing Stage	Total
COVID-19	112	30	142
Normal	112	30	142
Pneumonia	3,883	234	4,117
Normal	7,349	390	7,739
Total	11,456	684	12,140

7.3.2   DATA PREPROCESSING

With the intention of creating the majority of diminutive training samples, sudden increase in training data through quantity of arbitrary conversions hence CNN will not at all see twice exact same picture. This helps in preventing from overfitting that creates a better model. Keras libraries are mainly used in DL technique. So, in our novel method Keras library is utilized in preprocess the input images.

These are the pre-processing algorithms we built for our proposed model:

1. **Histogram Equalization:** Histogram equalization is one of the data pre-processing techniques mainly used to process input images. Also, this method utilized in IP techniques for contrast fine-tuning specified as bar chart. This approach typically enhanced worldwide dissimilarity of several images, particularly while usable image data is characterized by nearby dissimilarity values.

   i.   During this fine-tuning, the intensities of an image can be better scattered on bar chart. Hence it allows only less difference regions on images to achieve a superior dissimilarity.

   ii. This pre-processing algorithm achieved higher contrast next to scattering the most recurrent intensity values elsewhere. Moreover, this technique is functional in images with backgrounds and foregrounds which are either intelligent or shady.

   iii. The input is just a grayscale image and output is our histogram equalized image. OpenCV has a function to do this, cv2. equalizeHist().

2. **CLAHE (Contrast Limited Adaptive Histogram Equalization):** It is a variation of Adaptive histogram equalization which handles over amplification in contrast images. Also, CLAHE activate in small region denoted as tiles more willingly than whole region. The adjacent tiles are gathered using bilinear interpolation to eliminate artificial boundaries. The main important thing of using this algorithm is to enhance contrast on images.

7.4 CNN MODEL EXPLOITATION

Building DL model which is intended for learning distinguish among COVID-19 exaggerated X-ray and normal X-ray images. The basics of DL model especially CNN is used for distinguishing non-COVID and Coronavirus disease patient's mammogram images as explained in step-by-step in Table 7.4. In this proposed model, the author of this chapter have been taken three hidden layers in CNN where 224, 224 is the resize data, and it has three channels measured like our model size. Finally, flatten the features by using Σ (sigmoid) as to generate the output of NN (activation function) since we are having two-fold categorization problems, consequently production layer comprises of single-cell, adaptive moment estimation is an optimization algorithm perform along through Σ (sigmoid) function therefore reproduction compilation can be done together with cross binary entropy.

Model: "Sequential_3"

TABLE 7.4 How CNN Layers Used in Proposed Model?

Layers	Output Shape	Parameters
Conv2d_9	(None, 222, 222, 32)	896
Conv2d_10	(None, 220, 220, 128)	36,992
Max_pooling2d_7	(None, 110, 110, 128)	0
Dropout_9	(None, 110, 110, 128)	0
Conv2d_11	(None, 108, 108, 64)	73,792
Max_pooling2d_8	(None, 54, 54, 64)	0
Dropout_10	(None, 54, 54, 64)	0
Conv2d_12	(None, 52, 52, 128)	73,856
Max_pooling2d_9	(None, 26, 26, 128)	0

TABLE 7.4 *(Continued)*

Layers	Output Shape	Parameters
Dropout_11	(None, 26, 26, 128)	0
Flatten_3	(None, 86528)	0
Dense_5	(None, 64)	5,537,856
Dropout_12	(None, 64)	0
Dense_6	(None, 1)	65

Total parameters: 5,723,457
Trainable parameters: 5,723,457
Non-trainable parameters: 0
In addition to that, another three network models like Adam optimizer, RMSprop optimizer, and SGD optimizer have used to train the model, tuning the parameters till 10 epochs.

1. **Adam (Adaptive Moment Estimation) Optimizer:** The reason behind the name of Adam utilizes evaluation of first moment gradient (FMG) as well as second moment gradient (SMG) to modify the learning rate factor for every weight of NN. Adam optimizer is an adaptive learning rate model which calculates individual learning rates for several attributes. Also, Adam optimizer comprises of RMSprop along with SGD with impetus is being utilized for rising validation accuracy by performing fine-tuning parameters. The basic formula for Random Variable in Nth moment is defined as formula 1.

$$m_n = E\left[X^n\right]$$

where; X represents random variable whereas m specifies moment.

2. **RMSprop Optimizer:** It is an optimization algorithm which is equivalent as Gradient Descent Algorithm (GDA) along with momentum. Also, this optimizer controls the fluctuation in upright direction. As a result, we can have the ability to develop learning rate parameter in which it doesn't flows vertical direction that makes quicker convergence. The variation among RMSprop plus GDA is on what way gradients are estimated.

3. **Stochastic Gradient Descent (SGD) Optimizer:** To estimate the gradient which is for parameter modernization, we are using

weighted averages such as weight and bias dramatically. The major drawback of SGD optimizer is on owing to higher fluctuation we couldn't enhance the parameter learning rate hence it takes longer time to reach convergence results. Moreover, SGD along with momentum increase the convergence speed and fluctuation will be reduced.

7.5 DATA TRAINING

Initially, training data on the defined model without perform sheering, zooming, and horizontal shift hence the accuracy achieved nearly 50% which is attractive stumpy for real-time projects shown in Figure 7.3.

7.5.1 TRAIN THE DATA TO IMPROVE MODEL PERFORMANCE

For better performance of model by accuracy enhancement, we train the data through performing sheering, zooming, and horizontal shift. On one occasion, the input pictures (images) are enlarged, we have to modify the shape of the input image size as 224×224 along with batch size 32 leads to the training model as shown in Figure 7.3.

In splitting (specifically training) stage, I worked through 10 eras (epochs) with eight times for each era (epoch) done experimented with hyperparameters to produce better results.

```
3]: train_data = image.ImageDataGenerator(rescale = 1./255, horizontal_flip = True,)
    test_data = image.ImageDataGenerator(rescale=1./255)
```

```
0]: train_generator = train_data.flow_from_directory('CovidDataset/Train',target_size = (224,224),
                                          batch_size = 32, class_mode = 'binary')
    train_generator.class_indices

    Found 224 images belonging to 2 classes.

0]: {'Covid': 0, 'Normal': 1}
```

```
4]: validation_generator = test_data.flow_from_directory('CovidDataset/Val', target_size = (224,224),
                                           batch_size = 32, class_mode = 'binary')

    Found 60 images belonging to 2 classes.
```

FIGURE 7.3 Data training.

7.5.2 SUMMARIZING MODEL

The model has been evaluated to predict the performance in distinguish non-COVID mammogram images from coronavirus mammogram images. In addition to that, predicting validation loss as well as validation accuracy for improved understanding of requisite hyperparameters. The defined parameters achieved accuracy as 96.4% is well again but still need to increase accuracy rate by tuning hyperparameters.

7.6 RESULTS AND DISCUSSION

7.6.1 DATA ACQUISITION

The input image has been taken a a huge number of X-ray images of corona disease tested patients. We couldn't discover the disease affected patients but CNN can easily differentiate COVID-19 patients X-ray image from normal X-ray image. The image depicts that the left one indicates COVID-19 diagnosed patients the right side X-ray image indicates that the patient is fit (i.e.), patient is normal as well (Figure 7.4).

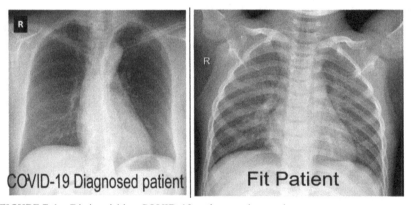

FIGURE 7.4 Distinguishing COVID-19 patients and normal.

7.6.2 DATA PRE-PROCESSING

Pre-processing technique is applicable to remove unnecessary features using histogram equalization algorithm, CLAHE algorithm, lung segmentation to segment the input images. Figure 7.5 illustrates that left image represent without undergone pre-processing and right image specifies after pre-processing.

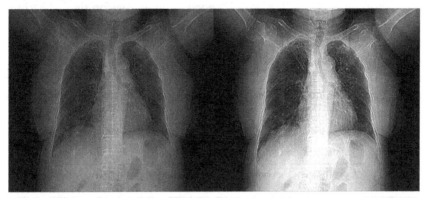

FIGURE 7.5   Distinguishing images before and after pre-processing.

7.6.3 MODEL EXPLOITATION

The model is deployed using several layers in the CNN approach by taking three hidden layers as input image size as (224, 224, 3) passes along convolutional 2D layer, max_pooling layer, dropout layer, flatten layer, dense layer, output layer to generate output as only one cell. The created model is depicted in Figure 7.6.

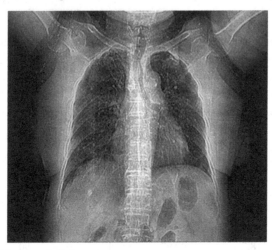

FIGURE 7.6   Model exploitation.

7.6.4 TRAINING DATA

Initially, data training has been evaluated through 10 eras (epochs) by means of eight steps for every epoch without performing sheering, zooming, and horizontal shift. In this stage, validation loss and validation accuracy mentioned in Table 7.5 estimated to calculate the performance of differentiate normal mammogram images from coronavirus mammogram and pneumonitis mammogram images. The validation accuracy achieved as 50% which is not better so tuning hyperparameters is proceeding in a later stage.

TABLE 7.5 Validation Loss and Accuracy – Stage I

Epoch	Loss (%)	Accuracy (%)	Validation Loss (%)	Validation Accuracy (%)
1/10	78.4	48.83	54.76	50
2/10	75.4	50.78	82.1	50
3/10	77.8	49.22	65.71	50
4/10	74.8	51.1	98.5	50
5/10	75.4	50.78	54.7	50
6/10	79.6	48.05	82.1	50
7/10	74.8	51.1	71.19	50
8/10	79.06	48.4	60.23	50
9/10	73.6	51.9	82.1	50

After tuning hyperparameters, the model has been trained through 10 eras (epochs) with eight times for every era (epochs) to achieve better outcomes illustrated in Table 7.6.

TABLE 7.6 Validation Loss and Accuracy – Stage II

Epoch	Loss (%)	Accuracy (%)	Validation Loss (%)	Validation Accuracy (%)
1/10	98.59	55.86	65.9	50
2/10	54.13	70.31	49.6	73.3
3/10	43.5	80.86	52.15	88.33
4/10	34.09	88.69	13.01	93.33
5/10	18.77	92.97	18.35	90

TABLE 7.6 *(Continued)*

Epoch	Loss (%)	Accuracy (%)	Validation Loss (%)	Validation Accuracy (%)
6/10	18.91	94.53	86.5	96.67
7/10	17.69	92.19	13.16	96.67
8/10	16.63	94.92	83.9	98.33
9/10	12.81	96.48	52.4	98.33
10/10	15.01	94.53	66.8	96.67

In our proposed model, the accuracy achieved as 96.4% in validation phase. Through accuracy estimation, the performance of distinguishing normal X-ray images and disease affected X-ray images. The greater accuracy leads to better model performance.

Figure 7.7 depicts the results of training accuracy as well as validation accuracy for 10 epochs using (224, 224, 3) as input image size via CNN.

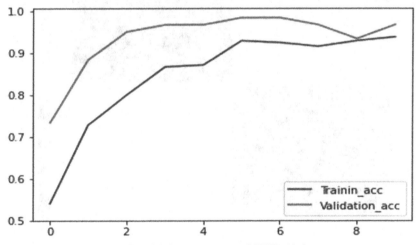

FIGURE 7.7 Training and validation accuracy on COVID-19 data.

7.7 METRICS EVALUATION

Confusion matrix describes the performance of classification model such as classifying normal images from Coronavirus patients images. To visualize the outcomes more comprehensible, confusion matrix is predicted for our model as follows in Table 7.7.

TABLE 7.7   Confusion Matrix – Proposed Method

	COVID-19	Normal
COVID-19	TP (true positive) must be high	FP (false positive) must be low
Normal	FN (false negative) must be low	TP (true negative) must be high

The following confusion matrix shown in Figure 7.8 is obtained by decoding the images, out of 30 COVID-19 affected patients, our experimental results obtained 30 people and zero incorrectly confidential, then among 28 regular (non-COVID) individuals, we obtained 28 patients as classified correct and 2 as classified wrongly.

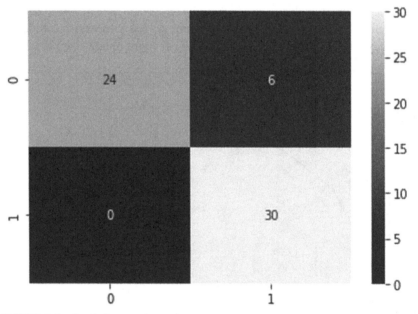

FIGURE 7.8   Confusion matrix.

Moreover, three CNN models such as Adam optimizer, RMS prop optimizer, and SGD optimizer were utilized for validation accuracy enhancement in categorizing COVID-19 and normal patients via mammogram images namely X-ray. The assessment of training accuracy and validation accuracy among these CNN approaches are evaluated in Table 7.8.

TABLE 7.8 Training and Accuracy Validation by Proposed Model

Epoch	CNN Model		Adam Optimizer		RMS Prop Optimizer		SGD Optimizer	
	Training Accuracy	Validation Accuracy	Training Accuracy	Validation Accuracy	Training Accuracy	Validation Accuracy	Training Accuracy	Validation Accuracy
1/10	50.96	50	56	50	44.2	50	52.25	58.3
2/10	49.58	50	70	83.3	52.5	50	63.1	81.67
3/10	51.31	50	80.6	93.3	51.6	50	68.74	81.67
4/10	50.64	50	86.7	95	60.2	58.3	84.18	86.67
5/10	53.64	50	88.07	98.3	65.39	91.67	77.23	68.33
6/10	50.59	50	95.1	96.6	80.3	50	74.34	83.3
7/10	50.83	50	96.7	96.6	69.32	88.3	83.54	90
8/10	50.77	50	92.7	96.6	87.7	91.6	85.94	53
9/10	46.59	50	91.17	96.6	88.7	95	75.58	**96.67**
10/10	50.83	**50**	94.8	**96.67**	90.6	**93.3**	91.49	93.3

The confusion matrix of these CNN models along with its validation accuracy scatter plot is described as follows. Initially, the implementation done using basic CNN model (noted in Figure 7.9) by estimating confusion matrix as well as scatter plot for training accuracy and validation accuracy.

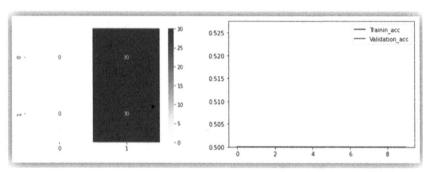

FIGURE 7.9   Scatter plot for confusion matrix with validation accuracy using basic CNN model after 10 epochs.

In Figure 7.10, the adaptive moment estimation (Adam) optimization algorithm deliberate purposely to train the NN model for categorizing COVID-19 patients from average individuals.

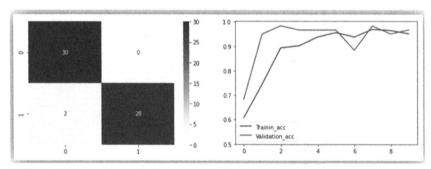

FIGURE 7.10   Adam optimizer model estimated confusion matrix with validation accuracy after 10 epochs.

Root mean square propagation (RMSprop) is also an optimization algorithm which necessitate for fine-tuning of hyperparameters to enhance the learning rate repeatedly. So, here we are using this algorithm for improvement in learning rate hyper-parameter exclusively which is to be usable while training phase of NN ranges between zero to one. Mainly this parameter organizes how fast the representation is modifying to the issues namely finding COVID-19. The confusion matrix was evaluated along with accuracy validation in detection of COVID-19 patients by implementing RMSprop as Figure 7.11 and SGD optimization algorithm as Figure 7.11 that makes faster convergence in disease diagnosis as depicted in Figure 7.12.

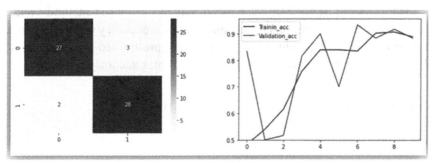

FIGURE 7.11 Confusion matrix with validation accuracy using RMSprop optimizer model after hyperparameters tuning.

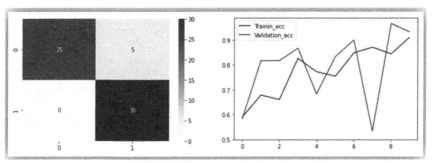

FIGURE 7.12 Confusion matrix with validation accuracy using SGD optimizer CNN model.

Moreover, comparison among pre-trained models and our proposed model have completed to highlight the performance in disease diagnosis. The comparison is mentioned in Table 7.9.

TABLE 7.9   Comparison Among Pre-Trained MODEL and Proposed

Pre-Trained Model	Training Accuracy	Validation Accuracy
Xception	92.19	88.94
VGG-19	89.86	88.14
InceptionV3	81.32	84.78
Proposed model	**94.8**	**96.67**

Hence, from Table 7.9, we noticed that our proposed method yields better outcomes in terms of training and validation accuracy when compared to pre-trained models. Figure 7.13 depicts the pre-trained models namely VGG-16, Inception, and Xception that determines accuracy with loss.

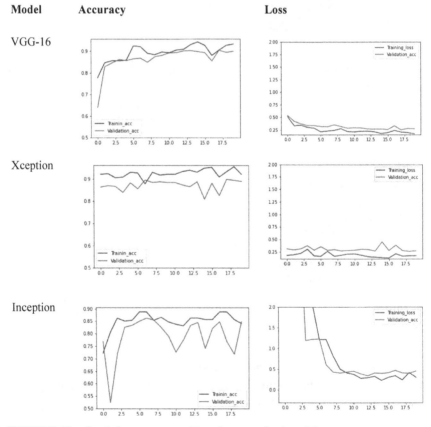

FIGURE 7.13   Graph for accuracy and loss for pre-trained models.

7.8 PERFORMANCE COMPARISON

Accuracy comparison for performance evaluation in detecting COVID-19 disease done by various researchers using DL techniques is shown in Table 7.10.

TABLE 7.10 Performance Comparison of Several Researchers in Detection on COVID-19 Using Various Deep Learning Techniques

Authors	Methodology Used	Accuracy Estimated	
Rahimzadeh et al. [18]	**XCeption**	**91.4%**	
	ResNet 50V2		
Xiang et al. [27]	Multivariate logistic regression	95.4%	
Irfan et al. [13]	LeNet	50%	
	VGG-16	70%	
	Inception	75%	
	ResNet50	84%	
Michael et al. [16]	VGG-19	Precision	X-ray – 86%
			US – 100%
			CT scan – 84%
Wang et al. [26]	Deep learning model	90%	
Javier et al. [14]	VGG-16	90%	
Muammer et al. [20]	CoviDetectioNet model	99.18%	
Amit et al. [6]	Deep CNN model	95.7%	
Harit et al. [11]	Deep learning technique	94%	
Moutaz et al. [19]	PA	94.8% – Australia	
	ARIMA	88.43% – Jordan	
	LSTM		
Rachna et al. [23]	Xception model	97.9%	
Harsh et al. [12]	nCovNet model	97%	
Ahmed et al. [4]	FO-marine predator algorithm	98.7%	
	Modified FO-MPA	97.2%	

Accuracy comparison in detection of COVID-19 disease using machine learning algorithms is depicted in Table 7.11.

TABLE 7.11 Performance Comparison of Several Researchers in Detection on COVID-19 Using Various Machine Learning Techniques

Authors	Machine Learning Algorithms Applied by Different Surveys	Accuracy
Ekta et al. [8]	Polynomial regression	93%
Elaziz et al. [9]	FrMEMS	96.09%
	Modified MRFO	98.09%

The surveys explained in Table 7.12 applied both machine learning and DL techniques for detecting COVID-19 patients from chest X-ray images.

TABLE 7.12 Detection of COVID-19 Disease Through Machine Learning and Deep Learning Techniques

Surveys	Methodology Used	Scope
Nariander et al. [21]	Machine learning – Support vector regression	Finding confirmed, death, and recovered cases separately applicable in real-time also helpful in medicinal diagnose.
	Deep learning – recurrent neural network, long-short-term memory	
Boran et al. [7]	Five machine learning algorithms – SVM, Naïve Bayes, LR, decision tree, and KNN CNN-ConvNet	Classifying X-ray images as normal, COVID-19, pneumonia disease

7.9 CONCLUSION

This chapter proposes the detection of COVID-19 disease and classifying images as normal and COVID-19 patents by CXR images. In accordance with appropriate medicinal research on signs of disease, we designed layers in CNN from CXR images which can rapidly finishes classification of coronavirus cases, regular, and pneumonitis patients. To achieve the above-mentioned goal, our proposal have been experimented several CXR images followed by pre-processing to remove irrelevant features, train the images for relevant features using 10 epochs leads to least accuracy. Moreover, this proposed model tried for hyperparameters tuning to achieve greater accuracy as 96.4%. This proposed experimental data demonstrates that our model has the capability to enhance accuracy in the detection of COVID-19 also classification of COVID-19, pneumonia, and normal via

CXR images. The future scope is parameters amendment using some other DL algorithm to achieve good performance in diagnose COVID-19.

KEYWORDS

- coronavirus
- first moment gradient
- gradient descent algorithm
- root mean square propagation
- second moment gradient
- stochastic gradient descent
- support vector machine
- World Health Organization

REFERENCES

1. Adedoyin, A. H., Ouns, B., Al-Turjman, F., & Moayad, A. (2020). AI techniques for COVID-19. *IEEE Access.* doi: 10.1109/ACCESS.2020.3007939.
2. Afshin, S., Marjane, K., Roohallah, A., Parisa, M., Ali, K., Delaram, S., Sadiq, H., et al., (2020). *Automated Detection and Forecasting of COVID-19 using Deep Learning Techniques: A Review.* arXiv:2007.10785v3 [cs.LG].
3. Ahmad, A., Sachin, A., Ishan, M., Patricia, B. M., Bina, J., & Xi, C. X., (2020). Artificial intelligence and machine learning to fight COVID-19. *Physiol. Genomics, 52,* 200–202. doi: 10.1152/physiolgenomics.00029.2020.
4. Ahmed, T. S., Dalia, Y., Ahmed, A. E., AlQaness, M. A. A., Robertas, D., & Mohamed, A. E., (2020). COVID19 image classification using deep features and fractionalorder marine predators algorithm. *Scientific Reports, 10,* 15364, https://doi.org/10.1038/s41598-020-71294-2.
5. Albahri, O. S., Zaidana, A. A., Albahri, A. S., Zaidana, B. B., Karrar, H. A., Al-Qaysi, Z. T., Alamoodi, A. H., et al., (2020). Systematic review of artificial intelligence techniques in the detection and classification of COVID-19 medical images in terms of evaluation and benchmarking: Taxonomy analysis, challenges, future solutions and methodological aspects. *Journal of Infection and Public Health, 13*(10), 1381–1396. https://doi.org/10.1016/j.jiph.2020.06.028.
6. Amit, K. D., Sayantani, G., Samiruddin, T., Rohit, D., Sachin, A., & Amlan, C. (2021). *Automatic COVID-19 Detection from X-ray Images Using Ensemble Learning with Convolutional Neural Network.* 1–14.

7. Boran, S., & Ilker, O., (2020). Detection of COVID-19 from chest X-ray images using convolutional neural networks. *SLAS Technology, 25*(6), 553–565. doi: 10.1177|2472630320958376.

8. Ekta, G., Ritika, J., Alankrit, G., & Uma, T., (2020). Regression analysis of COVID-19 using machine learning algorithms. *Proceedings of the International Conference on Smart Electronics and Communication (ICOSEC 2020).* IEEE Xplore Part Number: CFP20V90-ART; ISBN: 978-1-7281-5461-9.

9. Elaziz, M. A., Hosny, K. M., Salah, A., Darwish, M. M., Lu, S., & Sahlol, A. T., (2020). New machine learning method for image-based diagnosis of COVID-19. *PLoS One, 15*(6), e0235187. https://doi.org/ 10.1371/journal.pone.0235187.

10. Feng, S., Jun, W., Jun, S., Ziyan, W., Qian, W., Zhenyu, T., Kelei, H., et al., (2020). Review of artificial intelligence techniques in imaging data acquisition, segmentation, and diagnosis for COVID-19. *IEEE Reviews in Biomedical Engineering.*

11. Harit, A., Shubharthi, D., Bagish, C., Aishwarya, D., & Sanvinoth, P. S. S., (2020). Performance result for detection of COVID-19 using deep learning. *International Journal of Innovative Technology and Exploring Engineering (IJITEE)* (Vol. 9, No. 7). ISSN: 2278-3075.

12. Harsh, P., Guptaa, P. K., Mohammad, K. S., Morales-Menendez, R., & Vaishnavi, S. (2020). Application of deep learning for fast detection of COVID-19 in X-rays using nCOVnet. *Chaos, Solitons, and Fractals Nonlinear Science, and Non-Equilibrium and Complex Phenomena.* https://doi.org/10.1016/j.chaos.2020.109944.

13. Irfan, U. M., Syed, A. A. S., & Al-Khasawneh, M. A., (2020). A novel deep convolutional neural network model to monitor people following guidelines to avoid COVID-19. *Hindawi Journal of Sensors, 2020,* 15. Article ID: 8856801, https://doi. org/10.1155/2020/8856801.

14. Civit-Masot, J., Luna-Perejón, F., Manuel, D. M., & Anton, C., (2020). Deep learning system for COVID-19 diagnosis aid using X-ray pulmonary images. *Appl. Sci., 10,* 4640. doi: 10.3390/app10134640.

15. Jingxin, L., Lihui, Z., Yutong, Z., Zhong, Z., & Hairihan, W., (2020). Intelligent detection for CT image of COVID-19 using deep learning. In: *13ᵗʰ International Congress on Image and Signal Processing, BioMedical Engineering and Informatics (CISP-BMEI).*

16. Michael, J. H., Subrata, C., Manoranjan, P., Biswajeet, P., & Nagesh, S., (2020). COVID-19 detection through transfer learning using multimodal imaging data. *IEEE Access, 8,* Digital Object Identifier 10.1109/ACCESS.2020.3016780.

17. Mohammad (Behdad), J., Ali, L., Jakub, T., Zdeněk, P., Farimah, H., Pedram, L., Morteza, J., et al., (2020). Artificial intelligence and COVID-19: Deep learning approaches for diagnosis and treatment. *Special Section on Emerging Deep Learning Theories and Methods for Biomedical Engineering.* Digital Object Identifier 10.1109/ ACCESS.2020.3001973.

18. Mohammad, R., & Abolfazl, A., (2020). A modified deep convolutional neural network for detecting COVID-19 and pneumonia from chest X-ray images based on the concatenation of xception and ResNet50V2. *Informatics in Medicine Unlocked, 19,* 100360. https://doi.org/10.1016/j.imu.2020.100360.

19. Moutaz, A., Albara, A., Abdelwadood, M., Ajith, A., Vansh, J., & Salah, A., (2020). COVID-19 prediction and detection using deep learning. *International Journal of*

Computer Information Systems and Industrial Management Applications. ISSN 2150-7988 Volume 12 pp. 168–181.

20. Muammer, T. (2021). COVIDetectioNet: COVID-19 diagnosis system based on X-ray images using features selected from pre-learned deep features ensemble. *Applied Intelligence.* https://doi.org/10.1007/s10489-020-01888-w.

21. Narinder, S. P., Sanjay, K. S., & Sonali, A., (2020). *COVID-19 Epidemic Analysis using Machine Learning and Deep Learning Algorithms.* medRxiv preprint doi: https://doi.org/10.1101/2020.04.08.20057679.

22. Parul, A., Himanshu, K., & Bijaya, K. P. (2020). Prediction and analysis of COVID-19 positive cases using deep learning models: A descriptive case study of India. *Chaos, Solitons, and Fractals Nonlinear Science, and Nonequilibrium and Complex Phenomena.* https://doi.org/10.1016/j.chaos.2020.110017.

23. Rachna, J., Meenu, G., Soham, T., & Jude, H. D. (2020). Deep learning-based detection and analysis of COVID-19 on chest X-ray images. *Applied Intelligence.* https://doi.org/10.1007/s10489-020-01902-1.

24. Tengku, A. R., & Dadang, G., (2020). *IEEE 802.11ah Network Challenges Supports COVID-19 Prevention Team.* IEEE.

25. Wael, B. A., Amani, T., & Salma, H., (2020). Speech recognition for COVID-19 keywords Using Machine learning. *International Journal of Scientific Research in Computer Science and Engineering, 8*(4), 51–57.

26. Wang, S., Zha, Y., & Li, W., (2020). A fully automatic deep learning system for COVID-19 diagnostic and prognostic analysis. European Respiratory Journal. In press. https://doi.org/10.1183/13993003.00775-2020.

27. Xiang, B., Cong, F., Yu, Z., Song, B., Zaiyi, L., Qianlan, C., Yongchao, X., et al., (2020). *Predicting COVID-19 Malignant Progression with AI Techniques.* medRxiv preprint doi: https://doi.org/10.1101/2020.03.20.20037325.

28. Wang, L., Li, J., Wang, Y., Zhao, L., & Jiang, Q., (2010). Adsorption capability for Congo red on nanocrystal HneMFe204 (M=Mn, Fe, Co, Ni) spinel ferrites. *Chemical Engineering Journal.* doi: 10.1016/j.cej.-2011.10.088.

29. Somasekhara, R. M. C., Sivaramakrishna, L., & Varada, R. A., (2012). The use of an agricultural waste material, jujuba seeds for the removal of anionic dye (Congo red) from aqueous medium. *J. Hazard Mater., 203*, 118–127.

30. Sabnis, R. W., (2010). *Handbook of Biological Dyes and Stains: Synthesis and Industrial Applications* (pp. 106, 107). John Wiley & Sons, New Jersey, Canada, Springer.

31. Han, R., Ding, D., Xu, Y., Zou, W., Wang, Y., Li, Y., & Zou, L., (2008). Use of rice husk for the adsorption of Congo red from aqueous solution in column mode. *Bioresource Technol., 99*, 2938–2946.

32. Raymundo, A. S., Zanarotto, R., Belisario, M., Pereira, M. G., Ribeiro, J. N., & Ribeiro, A. V. F. N., (2010). Evaluation of sugarcane bagasse as bioadsorbent in the textile wastewater treatment contaminated with carcinogenic Congo red dye. *Brazilian Arch. Biology Technol., 53*, 931–938.

33. Hu, Q. H., Xu, Z. P., & Qiao, S. Z., (2007). A novel color removal adsorbent from hetero coagulation of cationic and anionic clays. *J. Colloid Interface Sci., 308*, 191–199.

34. Lian, L., Guo, L., & Wang, A., (2009). Use of CaCl$_2$ modified bentonite for removal of Congo red dye from aqueous solutions. *Desalination, 249*, 797–801.

35. Sanghi, R., & Bhattacharya, B., (2002). Review on decolorization of aqueous dye solutions by low cost adsorbents. *Color. Technol., 118*, 256–269.

36. Kumar, A., Kumar, S., Kumar, S., & Gupta, D. V., (2007). Adsorption of phenol and 4-nitrophenol on granular activated carbon in basal salt medium: Equilibrium and kinetics. *J. Hazard. Mater., 147*, 155–166.

37. Xin, X., Si, W., Yao, Z., Feng, R., Du, B., Yan, L., & Wei, Q., (2011). Adsorption of benzoic acid from aqueous solution by three kinds of modified bentonites. *J. Colloid Interface Sci., 359*, 499–504. https://www.sciencedirect.com/science/article/pii/S0304389407000052?via%3Dihub (accessed on 14 May 2022).

38. Erdem, B., Ozcan, & Ozcan, A. S., (2010). Adsorption and solid-phase extraction of 8-hydroxyquinoline from aqueous solutions by using natural bentonite. *Applied Surface Science, 256*, 5422–5427.

39. Ayari, F., Srasra, E., & Trabelsi-Ayadi, M., (2005). Characterization of bentonitic clays and their use as adsorbent. *Desalination, 185*, 391–397.

40. Gottipati, R., & Mishra, S., (2010). Process optimization of adsorption of Cr(VI) on activated carbons prepared from plant precursors by a two-level full factorial design. *Chemical Engineering Journal, 160*, 99–107.

41. Cronje, K. J., Chetty, K., Carsky, M., Sahu, J. N., & Meikap, B. C., (2011). Optimization of chromium(VI) sorption potential using developed activated carbon from sugarcane bagasse with chemical activation by zinc chloride. *Desalination, 275*, 276–284.

42. Myers, R. H., (1971). *Response Surface Methodology.* Allyn and Bacon, New York.

43. Box, G. E. P., & Hunter, J. S., (1957). Multi factor experimental designs for exploring response surfaces. *Ann. Math. Statist., 28*, 195–241.

44. Box, G. E. P., & Hunter, J. S., (1961). The 2K-p fractional factorial designs, parts I and II, *J. Technometrics, 3*, 311–458.

45. Muthukumar, M., Mohan, D., & Rajendran, M., (2003). Optimization of mix proportions of mineral aggregates using Box-Behnken design of experiments. *Cem. Concr. Compos., 25*, 751–758.

46. Benyounis, K. Y., Olabi, A. G., & Hashmi, M. S. J., (2005). Effect of laser welding parameters on the heat input and weld-bead profile. *J. Mater. Process. Technol., 164*, 978–985.

47. Tarangini, K., Kumar, A., Satpathy, G. R., & Sangal, V. K., (2009). Statistical optimization of process parameters for Cr (VI) biosorption onto mixed cultures of pseudomonas aeruginosa and *Bacillus subtilis. Clean, 37*, 319–327.

48. Box, G. E. P., & Hunter, J. S., (1957). Multi factor experimental designs for exploring response surfaces. *Ann. Math. Statist., 28*, 195–241.

49. Ozer, A., Gurbuz, G., Calimli, A., & Korbahti, B. K., (2008). Investigation of nickel (II) biosorption on *Enteromorpha prolifera*: Optimization using response surface analysis. *J. Hazard. Mater., 152*, 778–788.

CHAPTER 8

Breast Cancer Classification Using CNN Extracted Features: A Comprehensive Review

ARPIT KUMAR SHARMA,[1] AMITA NANDAL,[1] TODOR GANCHEV,[2] and ARVIND DHAKA[1]

[1]Department of Computer and Communication Engineering, Manipal University Jaipur, Rajasthan, India, E-mail: arvind.neomatrix@gmail.com (A. Dhaka)

[2]Technical University of Varna, Varna, Bulgaria

ABSTRACT

Medical imaging (MI) techniques have rapidly emerged as an important tool for clinicians. It has all the more as of late been utilized for preventive medication or evaluating for different illnesses like malignancy. It is especially pertinent that ordinary two-dimensional X-beams are not able to catch the imperfections. Thusly, an assortment of methods are utilized, contingent upon the speculated irregularity. Thus, tomography has to be further explored to improve breast cancer detection. The neural networks (NNs) help to predict the defect, which cannot be usually captured from conventional MI tools. Doctor insight of diagnosing and distinguishing bosom disease can be helped by utilizing some modernized highlights extraction and grouping calculations. This chapter presents an overview of different machine learning algorithms and examination between them which one is ideal to detect faster, and which are utilized to improve the exactness of foreseeing and predicting cancer.

Application of Deep Learning Methods in Healthcare and Medical Science.
Rohit Tanwar, PhD, Prashant Kumar, PhD, Malay Kumar, PhD, & Neha Nandal, PhD (Editors)
© 2023 Apple Academic Press, Inc. Co-published with CRC Press (Taylor & Francis)

8.1 INTRODUCTION

Breast malignant growth is profoundly extraordinary these days. Malignant development begins in cells and spreads throughout the individual. The development of circle cells develops a dominant part of tissue named as knot [1]. In this way, malignancy is significant. Amniocentesis underlying examination to disease masses are the most important indicator of serious cancer [14]. Masses are directed by the spaces perceived by wounds that can be raised by their primary development and a small amount of real estate. The second calcification pointer carries calcium leftovers. Minimal impressive spots in mammogram pictures. Clinical examination on breast malignant growth isn't novel yet nonattendance of legitimate techniques for early identification is as yet a test one [14].

As per the American Malignancy Society, Breast disease is the second most noteworthy reason for mortality among ladies in the USA. In 2013, more than 232,000 ladies will be recently determined to have Breast malignancy, and over 39,000 ladies will kick the bucket from this sickness [15]. Breast malignancy demise rates by and large increment with age. Around 95% of new cases and 97% of breast disease passing's happen in ladies 40 years old and more established. X-beam mammography was initially the solitary acknowledged imaging methodology for recognizing breast malignant growth. Be that as it may, this strategy isn't solid for imaging patients with thick breasts. Besides, it opens patients to ionizing radiations. As of late, new imaging methods have been utilized to supplement X-beam mammography and conquer a portion of its constraints and detriments. The American malignant growth society suggests that tolerance with a family background of breast malignant growth or who are more helpless against the illness because of different elements be screened every year starting age 40, or 10 years before the time of determination of a first-degree relative. When screening high danger ladies, the general location pace of mammography is 36% and the consolidated recognition pace of mammography and MRI is 92.7%. Identifying breast malignancy in high danger ladies utilizing mammography and US is 52%, contrasted, and 92.7% of consolidated mammography and MRI. Subsequently, the merged sufficiency for perceiving breast harmful development with mammography, US, and MRI is significantly higher than that of only one of these imaging methodologies. Right when the Breast threat is breaking down, a US or MRI-guided biopsy can assert the sickness. Overall, an ultrasound (US)-guided biopsy is supported considering its lower cost, relative straightforwardness, and a huger degree of patient comfort [16]. A couple of treatment choices are available after the assurance of breast dangerous development. The most notable treatment is breast mastectomy, which disposes of dangerous tissues and holds illness back from spreading [15]. Even after breast mastectomy, breast danger may rehash and still be a purpose behind death.

Breast malignancy is the second significant reason for death. After the cellular breakdown in the lungs breast, malignant growth happens the most. Breast disease happens ordinarily in ladies, infrequently in men. The insights show that 6% of ladies in India bite the dust because of breast disease. The sickness can be relieved if it is distinguished early. A lot of work has been done in this field to fix this deadliest infection. Malignancy starts in the cells, which are the fundamental structure squares of the tissues. Now and again, the development of cells turns out badly, and they don't kick the bucket as they ought to. At the point when this happens, a mass of cells creates, and they ultimately lead to malignancy. Clinical imaging has reformed the field in distinguishing malignancy. It isn't simply restricted to recognize malignancy; they can likewise distinguish different sorts of infection. As referenced before, breast disease can be relieved if it is identified before. Hereby an overview of different methods that can identify malignancy is introduced.

8.1.1 DEFINITION/KEY TERMS

As a foundation, the rudiments of bosom malignant growth and diagnostic tools are explained in this section.

1. **Cancer:** Malignancy is a strange advancement of cells that will in general repeat in an uncontrolled manner. Malignant growth is not a solitary sickness. It is a mix of over 100 extraordinary illnesses.
2. **Breast Cancer:** Disease is a crazy growth of bosom cells. Sorts of breast malignant growth breast disease start from various pieces. They begin from:
   i. Ductal disease;
   ii. Lobular disease;
   iii. Inflammatory bosom disease;
   iv. Paget sickness of the areola;

v.   Phyllodes tumor; and
vi. Angiosarcoma.

3. **Image Processing:** Picture preparing is a strategy to make an interpretation of a picture into its computerized structure and to get an improved picture or to remove a portion of the highlights; certain tasks are on the picture.

4. **Medical Image Processing:** Clinical imaging is the interaction or method used to address the inner design of the human body. Clinical imaging looks to reveal inside designs hidden by the skin and bones, just as to identify and treat illness.

5. **Medical Image:** Clinical pictures are the pictures that address the inner design of the human body. They are caught by utilizing different methods.

8.2   LITERATURE REVIEW

Vasantha et al. created channel strategies for playing the denoising cycle. The low-end channel method removes the noise but does not set the edges. Capture image details, however, accelerate the image. The advanced channel process also enhances image data. The positive response of the channel is disturbing to use both channel strategies to the fullest to achieve image quality [1]. Maintained by Jawad Nagi, a dynamic channel plan, measures a field with a square state. Channel generates the image smoother by filling in local image data without hiding the edge and maintaining the subtlety of images [2]. Juan Shah et al. developing a different way known as a mean channel. The incomprehensible of images with the unparalleled power of neighboring pixels. It turns out to be the best at killing Gausic noise [3]. Roselin et al. generated a histogram measurement strategy. This method processes mammogram imagery and improves the image quality minimum. Frequent pixel data can be fragmented [4]. Sanjay et al. create a cultivated process to improve the process. The local seed ensures space in a consistent environment. It may be heart-free and limited but depends on the pixel configuration request [5]. Indra et al. create a fusion process to distinguish between normal and abnormal areas in the chest muscles. The approach includes the calculation of the ASB and the growing area [6]. Dinsha made paper by splitting the breast tumor and doing something. In this strategy, the workload has been completed through the CLAHE process.

Using K-mean and fluffy c-mean, a separation cycle will be performed. Different exposures are removed from split images. Finally, the tests were performed using SVM and Bayesian classifiers [7]. Systematic seed selection calculations developed by Shan et al. In this figure, consideration is given to both spatial and spatial highlighting. You don't have to bother with any previous data or prepare a rating. What's worse is that the shaded areas with the lump can compare [8]. Ramani et al. is shown to separate the image by k-bunches. From a split image, every pixel is assigned to at least one group. By finding the distance between the pixel and between it, it is often added to the group. It collects a careful pixel based on a bunch for a small distance. Using a weight group, note the size of the mammogram image [9]. Ibrahim et al. developed a process called a black square level distribution plan. The most notable of these are the general, minimal, wide, medium, standard deviations, mean, defects, and mean total deviations [10]. Ibrahim et al. revived his previous process with the points to be chosen by selecting the main points [10]. Thus, it destroys the black foundation. Here, the key points chosen are homogeneity, a bunch of vague quality, and flexibility. Another beekeeping fraud strategy generated by Shanti et al. clears the noise from the photo, and the area of interest can be seen. Then a green c-conjunction is used, which means conjunction. Eighty-four prominent extracts contain fractal examination, correction, morphological, etc. [11]. Surendiran et al. developed an ANOVA DA proposal (Differential Analysis Discrimination). With the ANOVA DA scheme, a separate image classification was not considered. From major districts, 17 key points are developed based on the degree of determining whether the disease is benign or offensive. Highlight set can be reduced to a minimum and further enhanced by research techniques that are part of the head. This method improves the accuracy of the situation [12]. Kannan et al. created a molecular calculation of the molecule that supports the morphological calculation process. In this way, green standards are wrapped up and improve performance during pre-preparation. This gives an amazing result yet is identified by other planning techniques (Table 8.1) [13].

8.3　BASIC TECHNIQUES

There are different procedures utilized for identifying bosom malignant growth at the beginning phases. This segment depicts every strategy [17].

TABLE 8.1 A Literature Review with an Advantage

Author Name	Used Technique	Advantage
Vasantha et al. [1]	Channel strategies	Captures image details however accelerate the image
Jawad [2]	Dynamic channel plan	Channel generate the image smoother by filling in local image data without hiding the edge
Juan et al. [3]	Mean channel	Picture quality better
Roselin et al. [4]	Histogram measurement strategy	To improve the brightness of mammogram image
Sanjay et al. [5]	Cultivated process	The cultivated process to improve the process
Indra et al. [6]	Fusion process	Distinguish between normal and abnormal areas in the chest muscles
Surendiran et al. [12]	ANOVA DA proposal	Highlights set can be reduced to a minimum and further enhanced by research techniques that are part of the head.
Kannan et al. [13]	Molecular calculation	Green standards are wrapped up and improve performance during pre-preparation

8.3.1 MAMMOGRAPHY

In mammography, the X-beam is utilized as asymptomatic apparatus. Low-portion X-beams are utilized to see inside the bosom. It includes uncovering a piece of body to a little portion of radiation.

1. **Digital Mammography:** It is likewise called full-field computerized mammography (FFDM). Here the x-beam film is supplanted by gadgets that convert the x-beams into mammographic photos of the bosom. These pictures are inspected by the radiologist.

2. **Computer-Aided Detection:** This framework look in the digitized mammographic pictures for strange zones of thickness, the mass that may show the presence of malignancy.

3. **Breast Tomosynthesis:** It is additionally called the 3D mammogram and it is a high-level type of imaging where any number of pictures are caught from different points and they together structure a picture set.

8.3.2 THERMOGRAPHY

This technique is a non-ionizing strategy for examining inside design. The malignant growth is recognized by the warm markers in the infrared pictures. Infrared radiations are produced from the items which are over indisputably the zero temperature. The radiation is straightforwardly corresponding to the temperature. It increments as the temperature increments. The infrared pictures are caught by utilizing infrared cameras. The high temperature on the skin demonstrates the presence and level of malignancy. This technique is touchy to physiological changes that are the indications of disease or precancerous stage, which may prompt the development of tumor. This technique is a straightforward strategy to recognize malignant growth. The outcomes can be acquired rapidly with less spectator inclination. However, then again, the acquired outcome may not be exact, as they rely on the temperature of the body.

8.3.3 SONOGRAPHY

Sonography is a clinical imaging apparatus utilized for diagnosing or distinguishing bosom malignant growth. This method depends on the utilization of US. Ultrasound is a sound whose recurrence is more prominent than the human discernible recurrence (>20,000 Hz). The US pictures are made by sending beats of US into the tissues of the body. At that point, the sound echoes off the tissue, and they are recorded and changed over into a picture. Various tissues reflect the differing level of sound. A transducer is a hand-held gadget that is utilized to examine the bosom and show the progression of blood and the development of tissues. The US is utilized for seeing a few changes in the bosom that are not found in the mammogram. Ultrasound is the most ideal approach to see whether the irregularity is strong or liquid-filled. The Sonography strategy is more affordable than some other methods and they can distinguish tumors rapidly and without any problem.

8.3.4 DOPPLER ULTRASOUND

This strategy is utilized to screen the bloodstream in the bosom. It shows the blood course through the veins. The Doppler US waves are gone through the bosom and they reverberate from the tissues. These repeated waves are then changed over into a picture and showed on the screen. During Doppler US, a hand-held gadget called a transducer is ignored delicately the skin over the veins. It sends and gets enhanced sound waves. The sound waves ricochet off from the strong articles, including the platelets. The development of veins makes changes in the pitches of the sound waves. If there is no progression of blood, there won't be any pitch adjustment. By utilizing the reflected sound waves, a diagram or an image is produced and investigated for the location of the tumor.

8.3.5   COLOR DOPPLER

Color Doppler is a device that is not difficult to utilize and it is accessible in most ultrasonic imaging gadgets. It gives data about the vascularization in tissues. To show an exceptionally lethargic stream, it is important to apply insignificant tension on the bosom, consequently. It recognizes the squares in the progression of blood and blood course through the knots and consequently, the tumor can be distinguished early.

8.3.6   MAGNETIC RESONANCE IMAGING

This strategy comprises attractive field and radio recurrence beats. X-ray produces solid attractive beams into the body. It is utilized to test the inner organs, tissues, bones, and so forth the clinical pictures are shown on the PC screen and sent into an electronic sign and the subtleties are printed. This technique doesn't deliver any iron radiations and they recognize at which stage the disease is. They can be ready to catch the picture of both the bosom at the same time. It can undoubtedly distinguish any irregularities, tumor, and lymph hubs in armpits. It effectively identifies thick bosom tissues in more youthful ladies [18].

MRI is a significant instrument for neighborhood arranging before bosom disease medical procedure. Little intrusive malignancies and ductal carcinoma in situ can be identified utilizing bosom MRI because of amazing advances in transient goal and spatial goal. For high-hazard ladies, when supplemental screening is arranged, MRI is acted instead of US imaging. The American malignant growth society has refreshed its bosom disease imaging rules and now advocates bosom MRI for specific gatherings of high-hazard ladies. X-ray imaging utilizes an enormous magnet of 3–5 Tesla and RF loops to deliver 3D pictures of the bosom. The signs got are prepared to create the pictures. Contrasted with other imaging strategies, MRI is generally costly and requires an intravenous infusion of gadolinium, which causes the advancement of nephrogenic foundational fibrosis in a little gathering of patients with hindered renal capacity [17]. In this way, a patient with a background marked by renal infection will be unable to go through bosom MRI. Since this technique utilizes enormous magnets, it can't be performed for bosom malignant growth location in patients who have pacemakers or any metal inserts. X-ray imaging strategies are tedious and produce obscured pictures. Confounded MRI pictures necessitate those patients go through a similar imaging measure a few times [17]. Although MRI may save patients from a pointless medical procedure, there is a worry that discoveries on MRI may incite superfluous overabundance tissue expulsion or now and again superfluous mastectomy. At the point when sores are distinguished, MRI can be a dependable technique for biopsy or confinement of bosom disease. Many bosom biopsy frameworks are starting to arrive at the market, anyway they are not pervasive. From all the imaging strategies that have been explored, MRI has the most elevated affectability for distinguishing intrusive bosom carcinoma and can give significant data that isn't valued on mammograms. Bosom MRI screening is exceptionally reassuring when applied to high hazard gatherings.

8.3.7 POSITRON EMISSION TOMOGRAPHY

PET assists with identifying the region and cells that are influenced. The radiopharmaceutical called Fludeoxyglucose is infused first. It produces gamma beams. The FDG is recorded by the scanner and the pictures are built again and broke down. The dubious region is discovered where the signs are emphatically gathered.

8.4 SEGMENTATION TECHNIQUES

Division methodology could be arranged into brokenness and similitude in the picture. In brokenness, it very well may be grouped division. Comparability technique, focus on area-based division [18].

8.4.1   THRESHOLDING

It is a notable methodology. It portions pictures into closer view and foundation. In light of the exceptional, a paired picture can be made. This strategy endeavors to secure a force's esteem. The organized work is acclimated to the picture pixel force. This strategy is to part a picture into at least two sub-pictures by assessing with determined edge esteem G.

The pixel worth can be alluded to by f (I, j). Thresholding could be isolated into worldwide and neighborhood thresholding. Worldwide thresholding allotments the picture into two dependents on the earlier condition. At the point when G is consistent, at that point, the methodology is alluded to as worldwide thresholding. In neighborhood thresholding, segment a picture into sub pictures and determine the nearby properties of the picture. The downsides of this strategy are finding the edge esteem; they regularly do not have the affectability and particularity required for precise arrangement. It doesn't apply to multichannel pictures since it produces two classes in particular. The significant constraint is that there is a cover between dark levels of the article and foundation [18].

Methodology for choosing the breaking point esteem naturally:

- Select the principal estimation as far as possible;
- Separating picture dependent on cutoff esteem G. The outcome comprises two unique classes of pixels;
- T1 contains solid pixels $\geq$ G, while T2 contains pixels with estimations of <G;
- Calculate the strength esteems $\mu1$ and $\mu2$ for T1 and T2;
- G = 1/2 ($\mu1 + \mu2$);
- Perform the interaction until it varies from G.

8.4.2   EDGE-BASED SEGMENTATION

The edge can be addressed as a limit between areas with altogether different dim constructions. Generally, the edge happens between two distinct areas in the picture. Neighbor pressure changes in the picture are alluded to as edges. Adjoining force changes in a picture are alluded to as edges.

Two subsidiaries are accessible in edge administrators in particular: first request subsidiary administrators and second request subordinate administrators.

1. **Gray Histogram:** Dark level power and vigorously relies upon limit value. Tricky to locate the most extreme and least force since histogram isn't easy for the effect of commotion.
2. **Gradient-Based Method:** In the commotion in the picture, at that point angle based strategy functions admirably. This strategy includes fold-over slope administrators with the picture. Enormous estimation of the angle size is conceivable just when fast progress happens between locales. Slope-based few administrators and so on from the explanation, it is achieved that administrators generate outcomes contrasted with different administrators (Table 8.2).

TABLE 8.2 Edge Detection Techniques

Edge Detection Technique	Comparison of Techniques	
	Advantages	Disadvantages
Sobel	Smoothing edge	Commotion impact
Prewitt	Simple to discover neighborhood edge orientation	Not substantially more exact
Robert	Easy to plan	Affectability to commotion
Laplacian of Gaussian	Automatic scale selection	Reduces generally speaking differentiation in pictures
Zero cross	Smoothing edge	Tough to pick a limit
Canny	Wide scope of edges	Difficult to choose a conventional limit

Region-based strategy partitions a picture into a locale explicitly dependent on some pre-characterized models. This technique begins by choosing an arbitrary seed pixel and dissect it with adjoining pixels. Adjoining pixels that are comparable are assembled with the seed focuses. The thresholding technique amasses the limits among particular territories. A significant disadvantage is to choose the seed point physically. Beginning seed determination influences the eventual outcome. SRG is a straightforward, yet compelling calculation.

3. **Seeded Region Growing (SRG):** This count is quite obvious computation. The runs simply pick the test pixels that have a spot to gather. All separated region contains likeness and has no

withdrawn issue. Then again, picking a specific game plan of seed point may make particular division results. Furthermore, an immense issue is to understand the number of seed centers should be used around the grounds that a variety of pictures have fitting division number [18].

4. **Unseeded Region Growing (URG):** Unseeded region creating strategy subject to pixel resemblance incorporated by the territories. It does exclude any seed decision cycle. Thus, the computation creates the seeds. Unexpectedly, it has some invaluable advantages as sound and versatile [18].

   The accompanying URG contemplations:

   i.   The cycle begins with an R1 area that contains one-pixel;
   ii. Measure the current pixel esteem by a pixel estimation of R1;
   iii. If the estimation of the distinction <G (limit esteem), at that point added to the right spot, say Rj;
   iv. The pixel test isn't satisfied by structure; can be heard in another spot, named Ri;
   v.   If he isn't't happy with the two conditions, at that point he enters another spot, etc.

5. **Regional Classification and Integration:** Gap picture consistency and is addressed by quadtrees. In a quadtree, every hub has four ages and the root addresses the entire picture. The benefit of this strategy is to isolate the picture as per our application because of the level of breaking the conditions. The enormous draw is that it can make blocky parts. To keep away from the impeded part, the split is at a superior level.

8.5   FEATURES SELECTED FOR TAGGING BREAST CANCER IMAGES

To acquire the final image-wise classification, many patches are processed using a patch-wise classifier. Report of the institution the program is intended to guide radiologists and physicians who specialize in the decision-making process for breast cancer assists in patient management [14]. Breast density is an important criterion in the detection of breast cancer. The female breast is made of different types of tissues like fatty tissues, fibro glandular tissue, lymph node, milk duct, etc. Each tissue has its criteria of radiation absorption, due to which they can be categorized in the digital mammogram. As fatty tissues have low absorption of radiation, they look dark in digital mammograms as compared to fibro glandular tissue which is dense. Similarly, lymph nodes and milk ducts also look lighter as they are dense. Studies say that breast cancer is more common and undetectable if it occurs in women having denser breasts. The first tumor usually occurs in dense parts and second it is difficult to recognize easily due to similar intensity. So, studies say that if a computer can recognize the density of breast automatically, which is earlier done by a radiologist, chances of occurrence of breast cancer can be detected earlier and further screening can be done.

8.5.1 FEATURES BASED ON BREAST DENSITY

Figure 8.1 shows different density levels of breast tissues such as:

- Fatty tissues;
- Scattered density;
- Consistent density or heterogeneously dense;
- Extremely dense.

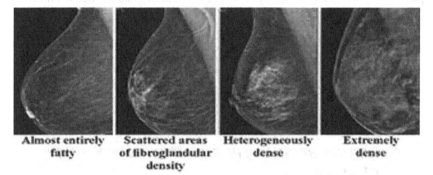

| Almost entirely fatty | Scattered areas of fibroglandular density | Heterogeneously dense | Extremely dense |

FIGURE 8.1 Image showing increasing level of breast density.
Source: https://breast360.org/topic/2017/01/01/breast-density-what-is-it-what-does-it-mean-to-me/.

8.5.2 FEATURE-BASED ON LUMP OR LESION TYPE

Abnormality in the breast is detected by the presence of a lump or lesion present in the breast region. However, it is further studied by the radiologist that whether this lump is cancerous or non-cancerous. This type of lump is divided into the following types.

- Benign lump or lesion;
- Malignant lump or lesion or tumor.

8.5.3 FEATURE-BASED ON A BENIGN LUMP OR LESION TYPE

Though all lumps or lesions present in the breast are not cancerous, the radiologist can find out through its discretion whether this tumor is benign or not. Though further clarification is done by biopsy of that particular tumor, most of the features can be detected through mammogram images. The benign tumor or lump can be of the following type.

- Fibroadenoma;
- Breast cysts.

8.5.4 FEATURE-BASED ON MASS OR LESION MARGIN

Mass or lesion margin is an important feature in detecting the type and probability of cancerous lump. It also tells radiologists about the nature of the expansion:

- Circumscribed;
- Micro lobulated;
- Obscured;
- Undefined;
- Speculated.

8.5.5 FEATURE-BASED ON THE DISTRIBUTION OF BREAST CALCIFICATION

Distribution modifiers are used to describe the calculation order. It helps in detecting the chances of the existence of a tumor:

- *Grouped* or *clustered* calcification;
- Linearly distributed calcifications;
- Segmentally distributed calcifications;
- Regionally distributed calcifications;
- Diffusely distributed calcifications.

8.5.6 *FEATURE-BASED ON LESION OR LUMP SITUATION*

- Round;
- Oval;
- Lobular;
- Unusual.

8.6 CONVOLUTIONAL NEURAL NETWORK FOR BREAST CANCER CLASSIFICATION

Assemble a calculation to consequently recognize if a patient is experiencing bosom disease by seeing biopsy pictures. The calculation must be very exact because the existences of individuals are in question. Figure 8.2 shows basic CNN architecture.

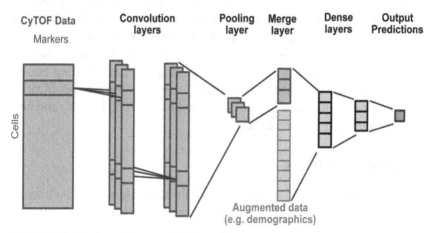

FIGURE 8.2   Basic architecture of CNN model.

8.6.1 *BASIC CNN ARCHITECTURE*

1. **Pooling:** The objective of this layer is to give spatial difference, which just implies that the framework will be equipped for perceiving an item in any event when its appearance changes somehow or on the other hand another.

2. **Fully Connected:** In a completely associated layer, we level the yield of the last convolution layer (CL) and interface each hub

of the current layer with different hubs of the following layer. Neurons in a related layer have a full relationship with all activations in the past layer, as found in customary NNs and work along these lines.

3. **Image Classification:** The overall process is as follows – our information is a preparation dataset that comprises N pictures, each named with one of 2 unique classes. At that point, we utilize this preparation set to prepare a classifier to realize what all of the classes resemble [14].

8.7 CONCLUSION AND FUTURE WORK

The current scenario, breast cancer classification using ML techniques make it easy and using its different type of methods and strategies is give better results and advantage so be shown in literature also. Review the major and minor types of strategies and methods with all parts according to methodology and algorithms. In the future researcher can do mix the two or three methods and a mathematical equation, make one, and find the best result same as with visualization for getting help to find out the accurate position of cancer and image. Researchers can do classification and detection as future of breast cancer and make a CNN, RNN, and help of other model try for best future scope.

ACKNOWLEDGMENTS

This work was supported by DST, New Delhi under India-Bulgaria International Scientific Cooperation, through project DST/INT/BLG/P-05/2019.

KEYWORDS

- convolutional neural network
- full-field computerized mammography
- machine-learning
- magnetic resonance imaging
- recurrent neural networks

REFERENCES

1. Vasantha, M., & Bharathi, V. S., (2011). Classification of Mammogram images using hybrid feature. *European Journal of Scientific Research, 57*(1), 87–96.
2. Nagi, J., (2010). Automated breast profile segmentation for ROI detection using digital mammograms. *IEEE EMBS Conference on Biomedical Engineering & Sciences.* Kuala Lumpur.
3. Shan, J., Ju, W., Yanhui, G., Zhang, L., & Cheng, H. D., (2013). Automated breast cancer detection and classification using ultrasound images: A survey. *Pattern Recognition, 43,* 299–317.
4. Thangavel, K., & Roselin, (2009). Mammogram mining with genetic optimization of anti-miner parameters. *International Journal of Recent Trends in Engineering, 2*(3).
5. Maitra, I. K., Nag, S., & Bandyopadhyay, S. K., (2011). Automated digital mammogram segmentation for detection of abnormal masses using binary homogeneity enhancement algorithm. *Indian Journal of Computer Science and Engineering, 2*(3).
6. Maitra, I. K., Nag, S., & Bandyopadhyay, S. K., (2011). Detection of abnormal masses using divide and conquer algorithm in digital mammogram. *International Journal of Emerging Sciences, 767–786.*
7. Dinsha, D., (2014). Breast tumor Segmentation and classification using SVM and Bayesian from thermogram images. *Unique Journal of Engineering and Advanced Sciences, 2,* 147–151.
8. Shan, J., Cheng, H. D., & Wang, Y., (2008). A Novel automatic seed selection algorithm for Breast ultrasound images in pattern recognition. *ICPR 19th International Conference.*
9. Ramani, R., Suthanthiravanitha, S., & Valarmathy, S., (2012). A survey of Current Image segmentation techniques for detection of breast cancer. *International Journal of Engineering Research and Application, 2*(5).
10. Ibrahim, S. I., Abuchaiba, & Elfarra, B. K., (2013). New feature extraction method for mammogram CAD diagnosis. *International Journal of Signal Processing, 6*(1).
11. Shanthi, S., & Bhaskaran, V. M., (2014). Modified artificial bee colony based feature selection: A new method in the application of mammogram image classification. *International Journal of Science, Engineering, and Technology Research, 3*(6).
12. Surendiran, B., & Vadivel, A., (2011). Feature selection using stepwise ANOVA discriminant analysis for mammogram mass classification. *ACEEE International Journal on Signal & Image Processing, 2*(1).
13. Rajiv, K. G., Marcus, K., & Kannan, S., (2010). Classification rule construction using particle swarm optimization algorithm for breast cancer datasets. *Signal Acquisition and Processing ICSAP International Conference, 233–237.*
14. Sharma, A. K., Nandal, A., Dhaka, A., & Rahul, D., (2020). A survey on machine learning-based brain retrieval algorithms in medical image analysis. *Health Technol., 10,* 1359–1373. doi: s12553-020-00471-0.
15. Sharma, A. K., Nandal, A., Dhaka, A., & Dixit, R., (2021). Medical image classification techniques and analysis using deep learning networks: A review. In: Patgiri, R., Biswas, A., & Roy, P., (eds.), *Health Informatics: A Computational Perspective in*

Healthcare: Studies in Computational Intelligence (Vol. 932), Springer, Singapore. doi: 978-981-15-9735-0_13.

16. Bhargava, N., Kumar, S. A., Kumar, A., & Rathoe, P. S., (2017). An adaptive method for edge-preserving denoising. In: *2nd International Conference on Communication and Electronics Systems (ICCES)* (pp. 600–604). Coimbatore. doi: 10.1109/CESYS.2017.8321149.

17. Gayathri, B. K., & Raajan, P., (2016). A survey of breast cancer detection based on image segmentation techniques. *International Conference on Computing Technologies and Intelligent Data Engineering (ICCTIDE'16)* (pp. 1–5). Kovilpatti, India. doi: 10.1109/ICCTIDE.2016.7725345.

18. Md. Islam, Naima, K., & Wen, H., (2013). A survey of medical imaging techniques used for breast cancer detection. *IEEE International Conference on Electro Information Technology*, 1–5. doi: 10.1109/EIT.2013.6632694.

19. Weblink: https://breast360.org/topic/2017/01/01/breast-density-what-is-it-what-does-it-mean-to-me/ (accessed on 14 May 2022).

CHAPTER 9

Multimodal Image Fusion with Segmentation for Detection of Brain Tumors Using a Deep Learning Algorithm

M. PADMA USHA,[1] G. KANNAN,[1] and M. RAMAMOORTHY[2]

[1]*Department of Electronics and Communication Engineering,*
B. S. Abdur Rahman Crescent Institute of Science and Technology,
Vandalur, Chennai, Tamil Nadu, India,
E-mail: padmausha@crescent.education (M. P. Usha)

[2]*Department of Artificial Intelligence and Machine Learning,*
Saveetha School of Engineering, SIMATS, Chennai, Tamil Nadu, India

ABSTRACT

Despite many years of research, brain tumors remain the deadliest and most life-threatening disease in all types of carcinoma (cancer). To support predictability in diagnosing brain tumors, a greater number of preclinical models needs to be produced for quicker and exact analysis. The proposed work utilizes the deep learning (DL) approaches to segment the brain tumor section using the fused multimodal brain tumor images, i.e., magnetic resonance imaging (MRI) and computed tomography (CT). This will enhance the integration of the various data streams that are available and makes the diagnosing process easier. Image fusion is performed by an averaging method and the tumor section is segmented applying by convolutional neural network (CNN) of the DL technique.

Application of Deep Learning Methods in Healthcare and Medical Science.
Rohit Tanwar, PhD, Prashant Kumar, PhD, Malay Kumar, PhD, & Neha Nandal, PhD (Editors)

Furthermore, image fusion of MRI images with CT has been reported to be useful in radio-surgery, neurosurgery, and postoperative treatment. The performance parameters such as structural similarity (SSIM) and Tumor size are calculated for image fusion and segmentation. SSIM is calculated for tissue and bone. The tumor stage is identified using tumor size.

9.1   INTRODUCTION

In recent decades, the prevalence of brain tumors has increased in people of all ages. The global mortality rate for cancers of the nervous system is estimated to be about 3.4 per 100,000. According to a study, there has been a significant increase in mortality of the elderly people of the age group especially 65–84 [15]. The mortality rate can be decreased and the survival rate can be increased for oncological patients with the help of the early and accurate diagnosis of the disease and also this is the major step of treatment. Existing treatment for cancer surgery, chemotherapy drugs, radiation therapy, or combining them. An increasing numerous of computer-aided design of neural imaging trained datasets are attempting to build the diagnostic tools that help in the visualization of the brain with MRI, automatic classification, and the diagnosis of disease [7].

Brain tumor detection can be performed by various artificial intelligence algorithms [11]. Among those commonly used are machine learning techniques such as random forest (RF), support vector machine (SVM), K-nearest neighbor (KNN) algorithm [5].

Even though the above machine learning techniques have several advantages, these are lagging in determining optimal parameters for non-linear separable data, produce poor accuracy and high classification error [1]. To overcome these issues, one of the commonly used deep-learning techniques called convolutional neural network (CNN) can be applied for brain tumor detection [2, 4]. The deep learning (DL) technique requires a large volume of dataset, whereas the machine learning technique does not require and it supports automated detection of a brain tumor [6]. RF, support vector machine (SVM), KNN algorithm are frequently used for pattern classification especially in tumor segmentation studies [12]. Among all, DL methods perform better tumor segmentation, classification, and prediction. To get better segmentation, preprocessing techniques may be adopted. A segmentation technique is then applied to the fused image. Preoperative patient planning, intraoperative dosage optimization, postoperative verification, and tumor response monitoring are all aided by image fusion [3]. When CT-MRI-based fusion is implemented region of tumor miss is minimal. Then the diagnosis and treatment can be determined according to that. The proposed techniques, pixel-based image fusion is performed and the CNN method is applied for segmentation. Pixel-based image fusion is having applications in various fields such as medical applications, remote sensing applications, video surveillance applications and photography [10].

9.2 EXPERIMENTAL METHODS

9.2.1 PROPOSED SYSTEM

The proposed system considers brain images and brain tumor images obtained from MRI and CT scan for fusion and segmentation processes. Brain slices are classified into Coronal slice, Sagittal slice, and axial slice. Axial slices are considered for the process since it covers all parts of the brain and it shows many tissues in a single slice.

For the proposed method analysis, MRI and CT images of the same slice of an axial plane are considered for analysis. As discussed in the literature survey performing image fusion before segmentation leads to the precision in the resulting image, which will aid the physicians to diagnose better of tumor part and the chances of missing it is lesser. Figure 9.1 shows the block diagram of the proposed method.

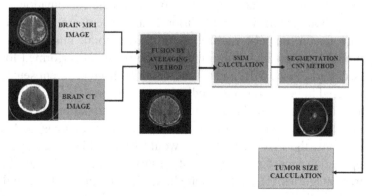

FIGURE 9.1 Proposed method of image fusion and segmentation.

9.2.2 PREPROCESSING

Medical image processing, unlike other forms of image analysis, is driven by very precise expectations. For example, we might want to detect tumors and recognize them. Alternatively, we may seek to assess natural or irregular structures. Pre-processing is needed before evaluating a medical image to refine the image. The goal here may be to boost contrast by removing any image artifacts or noise, or by highlighting distinctions between objects. For example, by improving the edges. The aim of preprocessing step is to remove noise from the input MRI and CT images. Also, preprocessing is performed to make the image suitable for further actual image analysis. In this work, preprocessing is performed using a median filter to remove salt and pepper noise which is more prevalent in biomedical scanned images. The median filter of mask size 3X3 is used after image capture.

9.2.3 IMAGE FUSION

Data from different pictures are merged in image fusion, enabling us to see a lot of detail in only one frame. The human mind is a perfect example of how an image fusion framework should be used [13]. Our eye will blend the images to show the finer details that are concealed in a single vision. The merged data aids in the creation of new images and incorporates all of the image's essential highlights. There has been considerable development in the area of multimodal image fusion, which has been increasingly important in several applications [16]. Fusion of images can occur in a variety of settings, including medical imaging (MI) applications, remote sensing, and machine vision. Image fusion allows the combination of two images without losing important information. There are several types of image fusion, namely, multi-view, multi-temporal, multi-focus, and multimodal among which multimodal image fusion is performed for the proposed work. Since MRI and CT are obtained from different sensors, it is considered to be multimodal medical image fusion.

There are so many techniques for fusing multimodal images, namely, Decision level, Feature level and Pixel-level fusion of images. Among those fusions, the Pixel-level fusion would be suitable for applying for this application according to the literature survey. Pixel-level image fusion has three categories: (i) maximum method, (ii) minimum method, and (iii) averaging method. Among this averaging method is chosen for fusing brain MRI/CT images. The averaging method is considering the average of corresponding pixels from the MRI and CT image in its respective position. The information from MRI and CT is combined into a single slice and this is passed onto for further image analysis such as segmentation of tumor region.

9.2.4 SEGMENTATION OF TUMOR REGION

There are many methods available for the segmentation of brain tumors. Conventional methods are thresholding, graph-cut method, etc. As an advancement to this, optimization of the segmented region can be performed by Particle Swarm Optimization (PSO) and Ant Colony Optimization (ACO) method. After drastic research in this area, AI methods have been initiated to implement in which machine learning for less volume of the dataset and DL for huge volume of a dataset. In the proposed method, CNN, a DL technique is implemented.

Figure 9.2 shows the sequence of CNN architecture. Convolutional layers are used to generate feature maps by using kernels to convolve a signal or an image [8]. The weights of the kernels bind the feature map unit that belongs to the previous layer. Backpropagation is used to adjust the kernel weights in the training process such that it improves the input characteristics [14]. CNN layers have minimum weights to train than thick fully connected (FC) layers, CNN is easier to train and less susceptible to overfitting since the kernels are shared among all units of the same feature maps.

Activation Function Rectifier Linear Units (ReLU) $f(y) = \max(0, y)$ (1)

The proposed method was validated with 12 pairs of imaging sets of MRI and CT, and this was downloaded from *Whole Brain Atlas from Harvard Medical School* [17]. The dataset contains both tumor-affected images and normal images.

9.3 RESULTS AND DISCUSSION

9.3.1 IMAGE FUSION AND SEGMENTATION

This section shows the experimental results of fused image and segmented image. The dataset contains totally 24 images in which 6 pair belongs to normal brain images and 6 pair belongs to brain tumor images. Images of size 512×512 is considered for analysis. Initially, images are median filtered with the mask of size 3×3. Performance of the proposed method was tested by using Structural Similarity(SSIM) for image fusion. Maximum value of Structural Similarity is 1. Higher the SSIM value-efficient is the fusion.

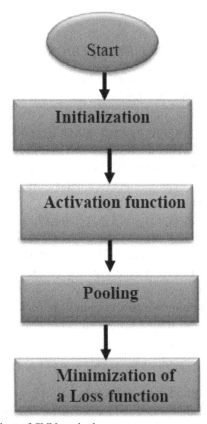

FIGURE 9.2   Flowchart of CNN method.

Output obtained from image fusion using averaging method is shown in Figure 9.3. Averaging method falls under Pixel-level fusion which provides more information for visual as well as data analysis. Later segmented results using the CNN method and the tumor size [9] are shown in Figure 9.4.

Input Image(MRI)   Input Image(CT)   Fused Image

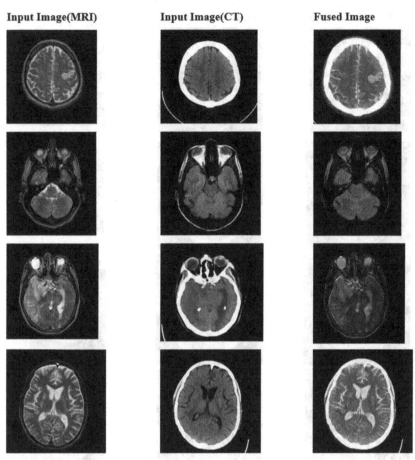

FIGURE 9.3   Image fusion using averaging method.

Source: Reprinted from Ref. [17]. © 1995–1999 Keith A. Johnson and J. Alex Becker.

Images considered in Figure 9.3 are normal MRI and CT brain images. Those images are fused by using the image averaging method in which corresponding pixels in both the images are taken average. From these fused images, Septum pellucidum, Precentral Gyrus, and Post Central Gyrus are visible combining the tissue as well as bone details.

Figure 9.4 shows the brain tumor affected MRI and CT images in which in both the modes, tumor region is visible and making them fuse hidden part of the tumor can be analyzed. The tumor region has been extracted with the support of CNN method.

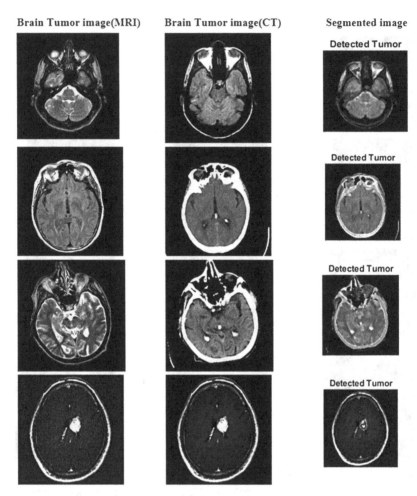

FIGURE 9.4   Segmentation of brain tumor using CNN method. © 1995–1999 Keith A. Johnson and J. Alex Becker.

9.3.2   *PERFORMANCE METRIC FOR IMAGE FUSION AND TUMOR SIZE FOR SEGMENTATION*

In this section, the performance metric of Image fusion is calculated and tabulated in Tables 9.1 and 9.2 and the tumor size of the segmented image is tabulated in Table 9.3.

The structural similarity index (SSIM) is a tool for assessing the visual effect of three image characteristics: structure, contrast, and luminance.

The SSIM is a quality measurement index that is dependent on the calculation of three terms: luminance, comparison, and structural. The aggregate index is a multiplication of the individual indices.

$$\text{SSIM}(x,y) = [m(a,b)]\alpha.[n(a,b)]\beta.[s(a,b)]\gamma$$

where;

$$l(a,b) = (2\mu_a\mu_b + C_1)/(\mu^2_a + \mu^2_b + C_1)$$

$$c(a,b) = (2\alpha_a\alpha_b + C_2)/(\alpha^2_a + \alpha^2_b + C_2)$$

$$s(a,b) = (\alpha_{ab} + C_3)/(\alpha_a\alpha_b + C_3)$$

where; μ_a is the local mean of image a; μ_b is the local mean of image b; σ_a is the standard deviation of image a; σ_b is the standard deviation of image b; and σ_{ab} is the cross-covariance for images a,b.

TABLE 9.1 Structural Similarity (SSIM) for Brain Tissue Images

	SET1	SET1	SET2	SET2	SET3	SET3
Method	SSIM	SSIM	SSIM	SSIM	SSIM	SSIM
	BONE	TISSUE	BONE	TISSUE	BONE	TISSUE
Average	0.5056	0.6249	0.5962	0.6377	0.5916	0.6291

TABLE 9.2 Structural Similarity (SSIM) for Brain Tumor Images

	SET1	SET1	SET2	SET2	SET3	SET3
Method	SSIM	SSIM	SSIM	SSIM	SSIM	SSIM
	BONE	TISSUE	BONE	TISSUE	BONE	TISSUE
Average	0.6031	0.6128	0.6417	0.6507	0.5033	0.6292

Figures 9.5 and 9.6 show the graphical variation of SSIM parameters for both brain images and brain tumor images. Tumor size is calculated for the segmented and the results are shown in Table 9.3.

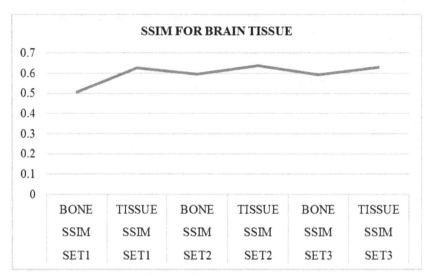

FIGURE 9.5   Structural similarity (SSIM) for brain images.

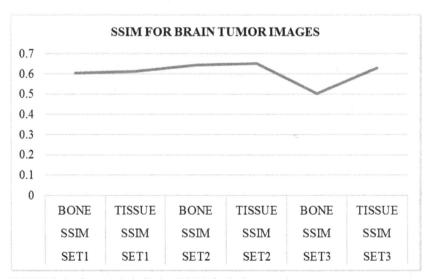

FIGURE 9.6   Structural similarity (SSIM) for brain tumor images.

TABLE 9.3   Tumor Size for Brain Tumor Images

Parameter	SET1	SET2	SET3	SET4
Tumor Size (in Pixels)	12.8516	4.57031	7.2265	17.8125

9.4 CONCLUSION

In cancer diagnosis and treatment, particularly, while applying postoperative radiotherapy, delineation of tumor cells is considered to be an important factor. The proposed framework aims to solve this issue by fusing MRI and CT images followed by the segmentation of the tumor region. This will help in increasing the probability of precise diagnosis. Geographical miss of the tumor while treating the tumor patient can be reduced by this proposed work. For image fusion, an averaging method is applied, which is a pixel-level fusion and this gives finer details in the fused image when compared with other image fusion techniques. Structural similarity (SSIM) is a performance metric that has been calculated and it was to be around 60%, which is a good result. After performing fusion, segmentation operation is performed with the technique CNN which gives the segmented tumor size as 4 to 13 pixels which will be then converted to a conventional unit.

KEYWORDS

- computed tomography
- convolutional neural network
- magnetic resonance imaging
- particle swarm optimization
- support vector machine

REFERENCES

1. Thahseen, P., & Anish, K. B., (2015). A deep belief network-based brain tumor detection in MRI images. *International Journal of Science and Research(IJSR), 6*, 495–500.
2. Ahmed, K., & Mahmoud, N., (2019). Classification of brain tumors using personalized deep belief networks on MR images: PDBN-MRI. *11ᵗʰ International Conference on Machine Vision, 11041.* doi: 10.1117/12.2522848.
3. Liu, S. F., Lu, J., Wang, H., Han, Y., Wang, D. F., Yang, L. L., Li, Z. X., & Hu, X. K., (2019). Computed tomography-magnetic resonance imaging fusion-guided iodine-125 seed implantation for single malignant brain tumor: Feasibility and safety. *Journal of Cancer Research and Therapeutics, 15*, 818–24.
4. Puneet, K. B., Akhil, K., Daleep, S., Mukesh, K. S., Satya, N., & Harvindra, S. K. (2015). Addition of magnetic resonance imaging to computed tomography-based

three-dimensional conformal radiotherapy planning for postoperative treatment of astrocytomas: Changes in tumor volume and isocenter shift. *South Asian Journal of Cancer, 4*. doi: 10.4103/2278-330X.149939.

5. Muhammad, W. N., Mohammed, A. A. G., Muzammil, H., Muhammad, A. K., Khalid, M. K., Sultan, H. A., & Suhail, A. B., (2020). Brain tumor analysis empowered with deep learning: A review, taxonomy, and future challenges. *Brain Sciences, 10*. doi: 10.3390/brainsci10020118.

6. Sérgio, P., Adriano, P., Victor, A., & Carlos, A. S., (2016). Brain tumor segmentation using convolutional neural networks in MRI images. *IEEE Transactions on Medical Imaging, 35*, pp. 1240–1251.

7. Ali, A., & Davut, H., (2018). Deep learning-based brain tumor classification and detection system. *Turkish Journal of Electrical Engineering & Computer Sciences, 26*. doi: 10.3906/elk-1801-8.

8. Aishwarya, N., Praveena, N. G., Rajalakshmi, B., Reshma, R., Rizwana, B. B., & Sowmy, D., (2020). Detection of brain tumor by image fusion based on convolutional neural network. *International Journal of Advanced Science and Technology, 29*, 6500–6509.

9. Adhi, L., & Thangadurai, A., (2012). Brain tumor segmentation and its area calculation in brain MR images using K-means clustering and fuzzy C-mean algorithm. *IEEE-International Conference on Advances in Engineering, Science, and Management*, 186–190.

10. Yu, L., Xun, C., Zengfu, W., Jane, W. Z., Rabab, K. W., & Xuesong, W., (2018). Deep learning for pixel-level image fusion: Recent advances and future prospects. *Information Fusion, 42*, 158–173.

11. Muhammad, O., Muhammad, A., Jiho, C., & Kang, R. P., (2019). Effective Diagnosis and treatment through content-based medical image retrieval (CBMIR) by using artificial intelligence. *Journal of Clinical Medicine, 8*. doi: 10.3390/jcm8040462.

12. Anish, V., Parth, D., & Shruti, J., (2020). Comparative analysis of various image fusion techniques for brain magnetic resonance images. *International Conference on Computational Intelligence and Data Science, 167*, 413–422.

13. Yann, L., L'Eon, B., Yoshua, B., & Patrick, H., (1998). Gradient-based learning applied to document recognition. *Proceedings of the IEEE, 86*, 2279–2324.

14. LeCun, Y., Boser, B., Denker, J. S., Henderson, D., Howard, R. E., Hubbard, W., & Jackel, L. D., (1989). Backpropagation applied to handwritten zip code recognition. *Neural Computation, 1*, 541–551.

15. Young, Z. K., Chae-Yong, K., Jaejoon, L., Kyoung, S. S., Jihae, L., Hyuk-Jin, O., Seok-Gu, K., Shin-Hyuk, K., Doo-Sik, K., Sung, H. K., et al., (2019). The Korean Society for neuro-oncology (KSNO) guideline for glioblastomas: Version 2018.01. *Brain Tumor Res Treat, 7*. doi: 10.14791/btrt.2019.7.e25.

16. Padma Usha, M., Kannan, G., Ramamoorthy, M., Sharmila, M., Huzaifa Anjum, G. A., Hairunnisha, M. S. H., (2020). Multimodal Brain Image Fusion using Graph Intelligence Method. International Journal of Research in Pharmaceutical Sciences, 11(2), doi:https://doi.org/10.26452/ijrps.v11i2.2293.

17. Johnson, K. A. and Becker, J. A. The Whole Brain Atlas. https://www.med.harvard.edu/aanlib/home.html

CHAPTER 10

Unrolling the COVID-19 Diagnostic Systems Driven by Deep Learning

SAKSHI AGGARWAL,[1] NAVJOT SINGH,[2] and K. K. MISHRA[1]

[1]Department of Computer Science and Engineering, Motilal Nehru National Institute of Technology Allahabad, Prayagraj, Uttar Pradesh, India, E-mail: sakshiaggarwal@mnnit.ac.in (S. Aggarwal)

[2]Department of Information Technology, Indian Institute of Information Technology Allahabad, Prayagraj, Uttar Pradesh, India

ABSTRACT

The novel coronavirus has shaken the entire Earth without being visible and making noise. The global response to this pandemic is commendable. In the whole scenario, whether it is a strategy formulation or reciprocation phase, we observed the influence of emerging technologies and ground-breaking artificial intelligence-based tools in pointing solutions to the healthcare community. There are plenty of diagnostic systems, attributed to COVID-19 pneumonia, leveraged deep-learning and data-driven frameworks. In the context of diagnostic steps, computer-aided diagnostic tools (CADT) continue assisting doctors and medical professionals for rapid diagnosis of new pneumonia. Digital imaging such as X-rays or CT scans aims to offer a clear picture of lungs demarcating the maligned and healthy regions. There is a never-ending list of such tremendous technological infrastructure providing relief to the people in the pandemic.

Our analysis is based on deep learning-driven diagnostic solutions for the healthcare society and global administration. The chapter initially

Application of Deep Learning Methods in Healthcare and Medical Science.
Rohit Tanwar, PhD, Prashant Kumar, PhD, Malay Kumar, PhD, & Neha Nandal, PhD (Editors)

covers three crucial keystones viz. *preventive measures, diagnostic measures, and immunization,* initiated across the nations. Remaining sections account proposed diagnostic models and their integration with the pandemic mitigation strategies. It also mentions some notable challenge that requires attention of the AI experts.

10.1 EPIDEMIOLOGY

10.1.1 GENESIS

Till December 2019, the world would have known the catastrophic impact of arms and ammunition, nuclear weapons, chemical reactors, or explosive devices onto human lives and how it could turn the wonderful planet Earth upside down. But then a diminutive virus has shown a disparate path of "destruction without being visible and making noise." This invisible enemy, initially appeared in Wuhan, capital of Hubei province in China, is named as Novel Coronavirus, or acronym for COVID-19 [1, 2]. Its origin is still a matter of contention, but consensus of early patients who had been exposed to this virus, had visited the Huanan Seafood Wholesale Market situated in Wuhan. The novel virus belongs to the family of bat coronaviruses and SARS-CoV (Severe Acute Respiratory Syndrome Coronavirus) and hence it is also called SARS-CoV2 [3]. The symptoms of this new viral disease, such as cough, headache, or body inflammation, seem closely related to any other regular pneumonia but the scenario is worst far than we thought.

SARS-Cov2 is a genus of single-stranded virus that affects the respiratory system of humans. In more severe situation, it enters into the nervous system that prevents proper functioning of the body. A patient can lose its ability to smell and taste, face trouble in breathing or control the balance over his body. It is highly contagious disease which keeps patient infectious for long weeks and the patient can shed the virus, even he is in recovery stage.

10.1.2 GLOBAL PLANNING AND RESPONSES

So far, each territory is attributed to at least a single case of COVID-19 disease [4]. Scientific researchers and the medical community across the world have collaborated to be familiar with the origin and pattern of the novel virus. Meanwhile, in the absence of suitable drugs and vaccines, short-term maneuvers [5] have been initiated by the countries to contain the virus spread. Active responses such as lockdown and night-curfews have become new normal. To lead a normal life but with restrictions, social-distancing proved effective in stabilizing the situation. In addition to it, administrative authorities focused on contact tracing and mass screening to prevent a region turning into an epicenter. Subsequently, patients with positive detection have to follow quarantine norms. They get isolated in order to break the chain of virus' fast replication and so does transmission.

In the whole scenario, whether it is a strategy formulation or recip-rocation phases, we observed the influence of emerging technologies [6] and ground-breaking artificial intelligence (AI)-based tools in pointing solutions to the healthcare community. The advance world of data science and digital platforms paved the way to keep the virus at bay. Moreover, dynamic technology of Big Data and Machine-Learning (ML) applications furnished maximum flexibility and regulation over containment measures. There are plenty of diagnostic systems [7–10], forecasting models [11, 12], or information-management systems [13], attributed to COVID-19 pneumonia, leveraged computational intelligence and data-driven frame-works. As public restrictions were imposed and people were totally in chaos, red-herring would have likely to spread out. But governing bodies made fruitful use of cloud-computing environments without sacrificing governance. The unprecedented digital tools provided agility to govern-ments for dissemination of information remotely. With the help of digital equipment or online media, people in isolated terrains are getting aware of government initiatives and programs instead of being carried away by panic attacks. Furthermore, in the context of diagnostic steps, computer-aided diagnostic tools (CADT) continue assisting doctors and medical professionals for rapid diagnosis of new pneumonia. Digital imaging such as X-rays [14] or computed tomography (CT) [15] scans aims to offer crystal-clear picture of lungs demarcating the maligned and healthy regions. It helps pulmonologist before he makes any decision and goes for further preparation. There is a never-ending list of such tremendous technological infrastructure providing resources and somehow relief to the people in the pandemic across the globe.

Inspired by this technological maneuver, we extend our deep insights about pandemic informatics via AI and deep-learning frameworks. Our analysis is based on deep learning (DL)-driven diagnostic solutions for the healthcare society and global administration. Despite, we also incorporate several biological and molecular-based activities that have been probably embraced by the medical fraternity in this crisis. The chapter initially covers three crucial keystones viz. *preventive measures, diagnostic measures, and immunization*, initiated across the nations. It accounts different proposed diagnostic models and their integration with the pandemic mitigation strategies.

10.2 PREVENTIVE MEASURES

While SARS-CoV-2 is covering a mile in a minute, collapsing the medical infrastructure, medical professionals and government authorities flag off several preventive steps to arrest the viral spread. In the absence of a particular vaccine and drug of this new pneumonia, many countries start implementing short-term measures, illustrated in Figure 10.1, enough to break the infection chain. Government of countries [16] such as Britain, India, Germany, and the USA began imposing lockdown over their citizens so that transmission rate could be controlled. This emergency lockdown protocol restricted individual's movement around the nation. Educational institutions were closed, and virtual classrooms through digital platforms were created for teachers and students. Owing to lockdown, international traveling was also affected. Domestic as well as international flights were temporarily suspended pertaining to strict protocol of staying home. In the flow of prevention, IT giants and multinational companies proffered a virtual workplace to the employees at their homes. This virtual workplace allowed them to sustain in the new-normal environment of work-from-home (WFH) [17].

Since COVID-19 is a highly contagious disease, social proximity was to be maintained. Even WHO [16] recommended in its guidelines the need of social-distancing or two-yard distance, being in the public premises. Along with that, wearing a mask was compulsory. Governments across the world urged people to wear masks and ensured the guidelines to be strictly followed. As an individual communicates frequently with his surrounding that contains humans, animals, or comes in contact with objects, hygiene becomes important and thus included in the response strategies by countries like US and India. According to WHO recommendation, hands must be washed thoroughly with soap and water for at least 15–20 seconds. The consumers can even use alcohol-based (at least 60% alcohol) hand sanitizers, if soap is out of reach.

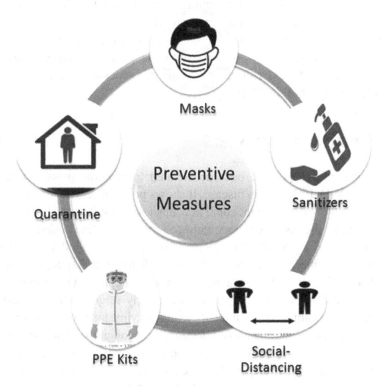

FIGURE 10.1   Different preventive measures were initiated to combat pandemic.

The frontline cluster including doctors, nurses, pharmacists, ambulance staff and remaining hospital personnel, was rigorously dealing with the COVID-19 positive patients not without carelessness. Safety measures such as personal protective equipment (PPE) kits had been distributed to protect health workers or other persons to defeat the infection. In general scenario, PPE kit is a standard way of avoiding infection, either borne through blood or air, while diagnosis or treatment. It is a kind of garment which consist of all important artifacts such as gown, gloves, masks, and face-shield. In this outbreak, it was successful and proven benevolent to medical practitioners who were in direct contact with the COVID-19 patient. However, in other secondary activities like medical-waste management, cleaning, burials, or other community-care concerned to the outbreak, PPE kits are extremely important.

Above mitigation strategies were successful in keeping low mortality rates of COVID-19 in several countries. Even after lockdown has been removed and there is ongoing research on vaccine or treatment, safety controls such as wearing masks, using hand sanitizers, PPE kits and following social proximity are dominant widespread. Government efforts can be seen as stringent in enforcing all the necessary guidelines with relaxations at certain intervals.

10.3   DIAGNOSTIC MEASURES

Diagnostic measures have been accelerated to promote mass testing of active COVID-19 infection disease. It shows positive results if a person is exposed to the virus or might come in contact with the infected person. He/ she should further take necessary steps to serve quarantine period or what-ever prescribed by the doctors. None of the testing method yields 100% accuracy, and the performance is highly variable. On different grounds, SARS-CoV-2 testing may not be that much effective. For instance, in asymptomatic individuals or person in the early stage of disease, results may fluctuate. Similarly, individuals with less viral load, or population of humans having no dominance of infection, are enough to get deviated from the results. Therefore, COVID-19 testing performance assessment is done through various parameters, such as the proportion of positive cases map to positive result and conversely proportion of negative cases map to negative result. In this course of action, we envisaged distinct features of testing methods, based upon Figure 10.2 [18], like RT-PCR and Antigen testing.

A virus is a minute creature of genetic material, enveloped in a molecule. Some virus particle contains ribonucleic acid (RNA) or some contains deoxyribonucleic acid (DNA). This genetic material or genome of the virus is detected through molecular reactions. The molecular approaches are often polymerase chain reaction tests, or PCR tests [19] and sequence of reactions rolling out. There is a dedicated laboratory to conduct these tests. In the PCR method, the first transformation of virus RNA to DNA takes place (i.e., reverse transcription), and then amplification is performed. The amplification runs multiple times to generate million copies of processed

DNA. It actually helps in sensing low volume of viral genetic particles in an individual's sample. That's why these molecular tests are generally sensitive.

In COVID-19 crisis, a variant of PCR, called RT-PCR (Reverse Transcription PCR) is adopted in many countries to support the testing resources. The real-time RT-PCR process is similar to conventional PCR technique, except the addition step of RNA-DNA conversion. While the efficacy of the test lies in the adaption for diagnostic step in early epidemics, such as Ebola [20] and Zika [21–23] virus or swine diseases, RT-PCR continues to deliver reliable outcomes within six or eight hours. It has been proven a robust system, not only for diagnostics, but also for studying and tracking the novel COVID-19 pathogen, which remains a challenge for researchers and practitioners.

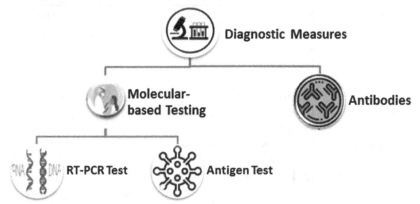

FIGURE 10.2 Diagnostic measures in terms of testing were promoted by the nations across globe.

On the contrary, another testing approach has been gaining momentum in COVID-19 pandemic. The Rapid Antigen Test (RAT) is based on detecting antigen molecules of virus in patient's swab sample. These molecules are basically blocks of proteins, lipids, or nucleic acids that have the potential of mimicking immune response. This test acts upon specific proteins of the virus, unlike the RT-PCR test which relies upon the genetic material. Generally, it is capable of interpreting active viral infection in an individual's body. A person with SARS-CoV-2 infection has spike antigens in his body and once the infection is gone, antigens vanish. It is a faster means of testing, and more susceptible to remote use and hence widely adopted for community testing.

As cases are surging, RT-PCR is not enough as it takes a longer duration to provide results. To alleviate this issue, RAT testing approach has been concurrently utilized for high true positive rate. Although RAT provides results in much lesser time relative to the RT-PCR test, but there are certain limitations append to this approach. The true negative results are much prone to errors and have to be verified with RT-PCR testing. Therefore, it should be viable only in containment regions or under the hospital's personnel supervisions.

TABLE 10.1 Comparison of Various Testing Methods Incorporated for Rapid COVID-19 Detection

	RT-PCR Test	RAT Test	Serological Test
Sample taken from…	Nasal or throat swab. Saliva is occasionally extracted.	Nasal or throat swab	Blood or finger
Detects the presence of…	Genetic material of the virus in the sample	Specific protein at the surface of the virus	Antibodies
How long process takes…	Eight hours or may be longer	15 to 30 minutes	1–3 days
How accurate the process is…	Highly accurate and does not require multiple cycles	Positive results need not to be confirmed while false-positive cases are also encountered. Negative results must be verified with the RT-PCR test	Sometimes it needs second test for more accuracy

In addition to diagnostic tests, the Food and Drug Administration of the US (FDA) [18] emphasizes for antibody test in terms of immunity check against COVID-19. We surely hear about immunity nowadays. Antibodies are the enzymes that are capable of fighting infectious diseases. They are secreted by our immune system when the body is prone to a threat, like bacteria or a virus. A stronger immune system means stronger protecting shield against infection. It take weeks for antibodies to develop when patient got infected and stays in blood for longer time even after recovery. For COVID-19 scenario, serological test is used to conduct antibodies test to determine if a patient is immune to novel coronavirus or not. In an alternative sense, if an immune system of a person has developed anti-bodies against COVID-19, he/she would have been exposed to the virus earlier at least once. To better understand the comparison between testing methods, Table 10.1 furnishes a clear picture.

10.4 IMMUNIZATION AND VACCINES

Immunization [24] is a vital pillar in the primary healthcare system and often regarded as a human right. It projects a powerful impact on millions of lives, saving across the countries. Vaccines, an indisputable model of immunization, provide shields to our body against harmful diseases and complement our natural immune system to stay stronger. If the body is introduced with a disease-prone micro-organisms, our innate system will immediately target them, fight off, and eventually destroy them. Vaccines stimulate the body's immune system by training and building it more effective. They have laid the foundation against national health menaces and will give vigorous ground to combat zoonotic diseases. In pre-COVID situations, distinguished vaccines have been discovered to save people of varying ages and defensive systems, from fatal diseases [25] such as tetanus, diphtheria, or measles.

The collaborative effort of scientists, enterprises, and comprehensive health organizations such as WHO, accelerating the COVID-19 crisis response by working day and night over more than 50 adequate vaccine candidates. It is the primary responsibility of health organizations to ensure the safety of vaccine. FDA does not compromise with safety factors, and hence there is a supplementary level of safety monitoring to swiftly stall the process, if anything goes wrong. The complete picture of clinical trials a vaccine or drug has to go through is illustrated in the given Figure 10.3. The SARS-CoV-2 vaccine [26] manufactures have been guided for full analysis and testing on the efficiency of potential vaccine on the grounds of adverse effects or critical parameters post-immunization. Human trials will have been given permission only when a vaccine is declared safe.

Once a vaccine is operational, allocation, and distribution will be the global aspect. COVAX, a joint venture initiated by WHO, GAVI [27] and CEPI [28], offered a window for global investment across vaccine candidates. It will be speeding-up COVID-19 vaccines manufacturing, availability, testing operations and treatment, along with the fair distribution across the globe.

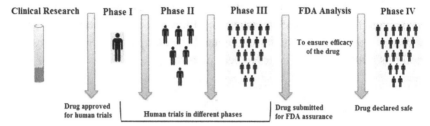

FIGURE 10.3 Different stages of vaccine in clinical trial.

10.5 ROLE OF DEEP-LEARNING IN COVID-19 DIAGNOSTIC SYSTEMS

The cooperation and collaboration of healthcare community and scientific researchers together in this phase is paramount important to the society. Their potential is not restricted to just foregrounding AI driven solutions but to handle gigantic data sets and think beyond uncertainties. With this affirmative stimulus, AI-based methods are adopted widely and integrated with mitigation strategies to accelerate the throughput. These methods are employed to support the aforementioned gold standard techniques to arrest SARS-CoV-2 around the globe. Moreover, DL (DL) paradigm has been seen in cases where there is a need to improve healthcare administration as well as diagnostics and prognosis. Usually, the conventional Machine Learning (ML) or data mining (DM) tools could not deal with inconsistencies arising in enormous data sets, unlike DL. It is a powerful subset of ML, that contain several layers of function responsible for resolving inconsistencies in data attributes. It is quite promising because it treats data in numerous manners, whereas conventional ML generally feeds on structured data.

In consequence with DL diagnostic models, radiography, and biomedical modalities have been suggested to be incorporated in the entire process of diagnosis and treatment. Various digital imaging techniques are blended with advance DL technology to build pathological applications that may be further used to distinguish COVID-19 and regular pneumonia. For each COVID-19 positive patient, chest X-rays (CXR) or CT scans are obtained in search of unexplored patterns attributed to novel pneumonia and could be served as supplementary data to DL frameworks. The medical imaging (MI) are valuable instruments to radiologists and physicians for identifying the affected regions of body and the extent to which tissues are deteriorated.

We have come across numerous applications of DL in developing SARS-CoV-2 predictive models in terms of diagnosis and prognosis. Hence, we aim to bring the best DL-based solutions forward associated with COVID-19 diagnostic systems that assist doctors and clinicians as well as uprooting the obstacles ahead of healthcare professionals. Consequently, the following sub-sections explore the latest biomedical image-based DL diagnostic models dedicated to the COVID-19 pneumonia according to patients' CT scan or X-ray as input images.

10.5.1 CT SCAN-BASED DIAGNOSTIC MODELS

Rapid detection of infection in lungs through CT scans attempts to augment existing healthcare planning for battling the pandemic. CT scans have a profound impact on organ's lesion segmentation and disease diagnosis. It is preferred over X-rays because of its multi-dimensional (specially three-dimension) view of the organ. Early signs of the disease could be identified from CT scans such as ground-glass opacity, or ground consolidation at the later stage. The qualitative information facilitated by CT scans thus makes it useful in COVID-19 situation. Some of positive COVID-19 CT scans are provided in Figure 10.4. However, it alleviates the cumbersome task of manual demarcating lung infections. Furthermore, extracting the annotations of infections is quite challenging for radiologists or clinicians, and often dependent upon individual's experiences.

Recently, burgeoning studies have been proffered leveraging CT scans and DL systems to screen out patients having infection in lungs caused by SARS-CoV-2 virus. For instance, location-attention based model [29] was proposed to determine infection rate due to COVID-19. Another DL-oriented health expert system was designed in Ref. [30] exploring three-dimensional (3-D) CT volumes owing to SARS-CoV-2 infection. Chen et al. [31] accumulated 46,096 CT screen slices including positive COVID-19 patients and other non-COVID diseases. In conjunction, modern DL frameworks such as U-Net++ [32] was trained by using CT volumes for diagnosing COVID-19 patients. The promising results are setting benchmarks for future systems and concurrently assisting radiologists in reference to medical crisis. Like U-Net++, other popular architectures have been deployed in COVID-19 diagnostic systems. It involves Resnet, incorporated in Refs. [29, 33], and U-Net [34], incorporated in Ref. [35]. Apart from diagnosis, DL technology is employed to delineate the contaminated regions from extracted quantitative features via CT scans, in order to assess severity [36] in the lungs, mass-screening [37], and quantifying lung infection [38–40] spread by COVID-19. The study in Ref. [41] allowed detection of pulmonary fibrosis with extensive accuracy. In Ref. [42], researchers came forward with a distinguished system which has the potential of identifying lung abnormality area and tissue-related disorders through CT scans.

TABLE 10.2   Few References for CT-based COVID-19 Diagnostic System

SL. No.	Data-Augmentation Used	Architecture Used	References
1.	–	UNet [44], DRUNET [45], FCN [46], SegNet [47], DeepLabv3 [48].	Zhang et al. [43]
2.	–	Inf-Net model that contains reverse attention modules and partial decoder	Fan et al. [15]
3.	–	UNet++	Zhou et al. [32]
4.	–	ResNet	Song et al. [33]
5.	Affine transformations, CGAN	AlexNet, VGGNet16, VGGNet19, GoogleNet, and ResNet50	Loey et al. [10]
6.	Annotations	UNet	Zhang et al. [30]
7.	Multi-scale spatial pyramid	Multi-scale CNN	Yan et al. [49]
8.	Size-balanced sampling	3D CNN	Ouyang et al. [50]
9.	Annotations	2D CNN	Bai et al. [51]
10.	Affine transformations	3D CNN	Xu et al. [52]

There is one limitation which is dominant in the above studies is, over-fitting where there is a paucity of required image-set happens. However, Kang et al. [53] tried to get rid of the problem by extending latent-representation learning to enforce compactness within COVID-19 and pneumonia acquired by community. The study suggested a unified model for latent representation that can encode different features derived from CT scans and finally project into class structures. Similarly, annotations-based weakly supervised learning method [30] was also tailored to alleviate the issue of overfitting. CT volumes and their 3D lung masks were considered as input to deep CNN [54, 55] which eventually outputs infection severity, further classified as COVID-positive or COVID-negative. In addition, the wide applications of data augmentation are also exploited with DL network architectures. Loey, Manogaran, and Khalifa [10] presented modern technology of conditional generative adversarial networks (CGAN) to subdue overfitting, enumerated by scarcity of CT radiography scans, as well as improve the network performance and COVID-19 detection rate. It used various deep CNN-driven network models viz. AlexNet [56], VGGNet16, VGGNet19 [57], GoogleNet [58], and ResNet50 [59]. Table 10.2 furnishes few CT-scan-based COVID-19 diagnostic systems with or without data augmentation technique.

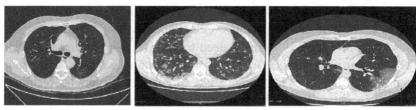

FIGURE 10.4   COVID-19 positive CT scans acquired from dataset.
Source: Reprinted from Ref. [78]. © X. Yang, X. He, J. Zhao, Y. Zhang, S. Zhang & P. Xie.

CT scans are crucial and informative radiography imaging technique adopted widely. It is helpful in the medical world where there is such an urgent need of automated diagnostic tools. It is considered as valuable signature which can reduce manual and cumbersome task of labeling the maligned or healthy regions by radiologists. Multilateral feature-representation is another highlight of this bioMI. The interpretation and deep understanding of CT scans with high accuracy can be performed with AI-based techniques such as DL. The comprehensive network architectures of DL such as ResNet, show promising results and could be further exploited for automatic analysis of radiological images. Although it has been widely embraced in digital diagnostic systems, we cannot overlook its demerits. CT scan machines are costlier and often not available in hospitals. Additionally, it has very high ionizing radiations, which can penetrate into the brain cells resulting the abnormal functionalities. Therefore, another radiography X-rays are preferred over CT volumes. There are numerous COVID-19 diagnostic studies which reveal the importance of CXRs.

10.5.2 X-RAY-BASED DIAGNOSTIC MODELS

To accelerate the mass-screening of COVID-19 pneumonia, several DL models based on CXRs have been proposed. Figure 10.5 contains some X-ray showing positive signs of COVID-19. Rahimzadeh and Attar [60] developed an automatic classification of COVID-19 patients based on advance CNN architecture of Xception and ResNet50V2. Alqudah et al. [61] exploited X-rays in developing computer-aided diagnostic tool for COVID-19 classification. It embedded different classifiers such as Support Vector Machine (SVM), Random Forest (RF) and CNN. Besides it, generative Adversarial Network (GAN) was also propounded along with astonishing networks such as GoogleNet, AlexNet, and ResNet18. Fan, Chen, Chen, and Dong [62] used X-rays as input images for COVID-19 diagnosis system leveraging DL frameworks. Going ahead with the list, Apostolopoulos and Mpesiana [9] utilized deep transfer learning mechanism with CNN so that necessary COVID pneumonia features could be drawn out from X-rays. Different DL variants viz. MobileNet, Inception, Xception, VGG19, Inception-ResNetV2 are incorporated to predict COVID-19 cases. In the proposed work of Bandyopadhyay and Dutta [63], LSTM-GRU was employed, not only for diagnosis, but to classify deceased, confirmed, discharged, or death cases of COVID-19 after rigorous analysis of X-rays. Similarly, Chen et al. [64] introduced DL-oriented COVID-19 detection system using deep interpretation of X-rays.

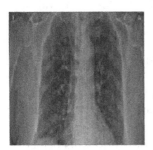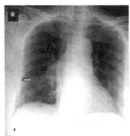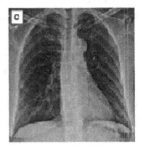

FIGURE 10.5   COVID-19 positive chest X-rays taken from dataset.

Source: Reprinted from Ref. [79]. © Joseph Paul Cohen, Paul Morrison, Lan Dao

Another feature extraction and representation-driven strategy was exploited in the proposed work of Kumar and Kumari [65], coupled with powerful SVM as a classifier to predict COVID-19 patients. ResNet50, combined with SVM, could be considered as a statistically and qualitatively superior to various models. Horry et al. [66] came up with a scheme that emphasizes pre-trained network architecture for COVID-19 detection from X-rays. The comprehensive analysis of Hemdan et al. [67] involved seven advance architectures of DL such as VGG19, InceptionV3, MobileNetV2, DenseNet201, ResNetV2, InceptionResNetV2, and Xception, to deal with COVID-19 X-rays. Furthermore, an integrated environment of CNN-LSTM was employed for X-rays by Islam, Islam, and Asraf [7] to detect novel Coronavirus. The study proposed by Brunese et al. [68] offered a distinct DL-driven detection system which aimed to differentiate COVID-19 pneumonia and other regular pulmonary diseases. Later on, it identified region-of-interest attributed to COVID-19 in CXRs.

TABLE 10.3 References for X-Ray-Based COVID-19 Diagnostic System

SL. No.	Data Augmentation Used	Architecture Used	References
1.	–	VGG-16 [57]	Brunese et al. [68]
2.	Affine transformations	CNN, LSTM	Islam et al. [7]
3.	Offline augmentation	Bayes-SqueezeNet	Ucar et al. [69]
4.	Affine transformations	CNN	Wang et al. [70]
5.	Affine transformations	CNN	Nour et al. [71]
6.	–	CNN	Das et al. [72]
7.	–	CNN	Ozturk et al. [73]
8.	–	ResNet18	Oh et al. [74]
9.	–	CNN	Rajaraman et al. [75]
10.	One-hot encoding [76]	VGG19, MobileNet, DenseNet121, InceptionV3, ResNetV2, Inception-ResNet-V2, Xception, MobileNetV2,	Hemdan et al. [67]

Like CT scans, X-rays are also scarce, in the context of academic research and public accessibility for COVID-19 analysis. However, researchers [77] presented a unique approach of fuzzy induction rule and composite Monte-Carlo methodology to address overfitting. Conventional data augmentation techniques were also applied in studies like Nour, Comert, and Polat [71]. Another methodology [74] was foregrounded dealing with an insignificant number of CXRs. The encouraging model was driven by patch-based CNN that tuned on few parameters for COVID-19 forecasting. Tables 10.3 give some references for X-ray-based COVID-19 diagnostic model either utilizing augmentation technique or simply going with DL architectures.

X-rays are more emphasized in encouraging studies due to its economic accessibility and intuitive implementation. Unlike CT scans, the human's body has to tolerate less radiations. It proves fruitful in segmenting potential area-of-interest before proceeding for surgery. Even before the pandemic, radiologists have been using X-rays for detecting fractures, internal injuries, or any kind of abnormalities. In spite its acceptance, X-rays are not without limitations. Rib-cage around lungs absorbs radiation which affect diagnostic data. Microscopic components are difficult to capture through X-rays as it cannot strongly interact. Lastly, X-rays does not furnish 3D view of organs as CT scan does.

10.6 CHALLENGES WITH DEEP APPROACHES

Thanks to the DL paradigms, it have surely enhanced the segmentation performance and managed complicated conditions with more steadiness. However, there are some challenges that need to be reviewed while exploiting DL frameworks in practice. First of all, consolidating medical annotations is a tough task for AI experts and analysts as DL network architectures like VGGNet, ResNet, requires a significant number of annotations to perform training task. Annotating samples demand clinical experiences and are often considered as tedious and costly. Authentic COVID-19 annotations are available through limited in number. Secondly, overfitting in deep models is prevalent in the studies. As mentioned before, it is a vital issue that caused by unavailability of instances required for the problem. Typically, it arises due to unreasonable size of training set. A third major challenge is concerned with minimizing network training time and simultaneously faster convergence. And finally, resource allocation and hardware requirements for deploying DL-based services, are some other limiting factors that could not be avoided. Generally, DL architectures are heavy-weight and consume larger workspace.

Therefore, the aforementioned challenges need to be addressed although there are retrospective studies showing encouraging results. DL models have great potential and reached to greater heights, but looking at varying aspects, it should not be confined to academic research but be flourished in practice on a large scale. It perhaps induces the atmosphere of scientific researchers and healthcare professionals, working side-by-side.

10.7 PUTTING ALL TOGETHER

With this compilation, the conceptual phenomena and platforms composed by AI-based schemes, especially inquisitive Deep-Learning technology, and adequate to deal with COVID-19 pandemic have been emphasized. Initially, when Coronavirus was in full swing, there were unprecedented challenges before global administration as well as the medical community. Being unfamiliar with this new micro-organism, nations have come together over a single platform to contain the virus spread and prevent it from taking over humans' life. Universal responses and planning were incredible. Decision-making strategies were rapidly induced in the emergency medical crisis in terms of preventive measures, diagnostic measures, and immunization or vaccination, by the governments around the world.

Besides, it is also important to mention the role of DL-driven diagnostic frameworks contributed to COVID-19 pneumonia detection. Various architectures such as CNN, LSTM, and CGAN were developed that have been discussed in this chapter. Distinguishing high-risk patients, recognizing contaminated areas in the lungs, evaluating risk-severity, and radiology are the main issues for which various proposed models are comprehensively dealing with. These AI-based prediction tools aid radiologists and practitioners in real-time by analyzing enormous image data, applying techniques or high-level intelligence, training machines, with the motive of fast diagnosis and improved performance. However, considering the limitations and challenges of DL-based approaches applicable in COVID-19 scenario, deeper knowledge and understanding is yet to be accomplished. The acquired success in conquering COVID-19 is extremely dependent on data-driven frameworks, enterprises that offer advanced platforms, and paradigms to achieve objectives, realize more lives, and get prepared for future crises.

KEYWORDS

- artificial intelligence
- computed tomography
- computer-aided diagnostic tools
- data mining
- deep learning
- machine-learning
- personal protective equipment
- rapid antigen test

REFERENCES

1. Munster, et al., (2020). A novel coronavirus emerging in China-key questions for impact assessment. *N. Engl. J. Med., 382,* 692–694.
2. Wang, C., Horby, P. W., Hayden, F. G., & Gao, G. F., (2020). A novel coronavirus outbreak of global health concern. *Lancet, 395,* 470–473.
3. UNICEF.org   Homepage.   https://www.unicef.org/india/coronavirus/covid-19 (accessed on 14 May 2022).
4. Worldometers (2020). https://www.worldometers.info/coronavirus (accessed on 14 May 2022).
5. Yasir, A., Hu, X., et al., (2020). Modeling impact of word of mouth and e-government on online social presence during COVID-19 outbreak: A multi-mediation approach. *Int. J. Environ. Res. Public Health, 17,* 2954. doi: 10.3390/ijerph17082954.
6. Sera, W., Mamas, A. M., et al., (2020). Applications of digital technology in COVID-19 pandemic planning and response. *Lancet, 2*(8), E435–E440. doi.org/10.1016/ S2589-7500(20)30142-4.
7. Islam, M. Z., Islam, M. M., & Asraf, A., (2020). A combined deep CNN-LSTM network for the detection of novel coronavirus (COVID-19) using X-ray images. *Informatics in Medicine Unlocked.* doi.org/10.1016/j.imu.2020.100412.
8. Butt, C., Gill, J., Chun, D., & Babu, B. A., (2020). Deep learning system to screen coronavirus disease 2019 pneumonia. *Applied Intelligence.* doi.org/10.1007/ s10489-020-01714-3.
9. Apostolopoulos, I. D., & Mpesiana, T. A., (2020). COVID-19: Automatic detection from X-ray images utilizing transfer learning with convolutional neural networks. *Physical and Engineering Sciences in Medicine.* doi.org/10.1007/s13246-020-00865-4.
10. Loey, M., Manogaran, G., & Khalifa, N. E. M., (2020). A deep transfer learning model with classical data augmentation and CGAN to detect COVID-19 from chest

CT radiography digital images. *Neural Computing and Applications*. doi.org/10.1007/s00521-020-05437-x.

11. Arora, P., Kumar, H., & Panigrahi, B. K., (2020). Prediction and analysis of COVID-19 positive cases using deep learning models: A descriptive case study of India. *Chaos, Solitons, and Fractals*. doi.org/10.1016/j.chaos.2020.110017.

12. Zeroual, A., Harrou, F., Dairi, A., & Sun, Y., (2020). Deep learning methods for forecasting COVID-19 time-series data: A Comparative Study. *Chaos, Solitons, and Fractals*. doi.org/10.1016/j.chaos.2020.110121.

13. N. I. Centre, (2020). Aarogya Setu mobile app. [online available]: www.mygov.in/aarogya-setu-app (accessed on 14 May 2022).

14. Mangal, A., et al., (2020). *CovidAID: COVID-19 Detection Using Chest X-ray*. arXiv.

15. Fan, D. P., et al., (2020). *Inf-Net: Automatic COVID-19 Lung Infection Segmentation from CT Scans*. arXiv.

16. World Health Organization. Homepage www.who.int (accessed on 14 May 2022).

17. Stadtlander, L., Sickel, A., Civita, L. L., & Giles, M., (2017). Home as workplace: A qualitative case study of online faculty using photovoice. *Journal of Educational Research and Practice*. doi: 10.5590/JERAP.2017.07.1.04.

18. US Food and Drug Administration Homepage www.fda.gov (accessed on 14 May 2022).

19. Grover, N., (2020). PCR, antigen, and antibody: Five things to know about coronavirus tests. *Horizon: The EU Research and Innovation Magazine*.

20. CDC, (2016). *Outbreaks Chronology: Ebola Virus Disease*.

21. He, S., et al., (2015). Repurposing of the antihistamine chlorcyclizine and related compounds for treatment of hepatitis C virus infection. *Science Translational Medicine*.

22. Barrows, N. J., et al., (2016). A screen of FDA-approved drugs for inhibitors of Zika virus infection. *Cell Host Microbe*.

23. Xu, M., et al., (2016). Identification of small-molecule inhibitors of Zika virus infection and induced neural cell death via a drug repurposing screen. *Nature Medicine*.

24. Centers for Disease Control and Prevention, (2021). CDC, Vaccines and Immunization Homepage. https://www.cdc.gov/vaccines/index.html (accessed on 24 May 2022).

25. Centers for Disease Control and Prevention, (2021). CDC, Vaccines and Preventable Diseases Homepage. https://www.cdc.gov/vaccines/vpd/vaccines-diseases.html (accessed on 24 May 2022).

26. Rawat, K., Kumari, P., & Saha, L., (2020). COVID-19 vaccine: A recent update in pipeline vaccines, their design and development strategies. *European Journal of Pharmacology*. doi.org/10.1016/j.ejphar.2020.173751.

27. GAVI Homepage. www.gavi.org (accessed on 14 May 2022).

28. CEPI Homepage. www.cepi.net (accessed on 14 May 2022).

29. Xu, X., Jiang, X., et al., (2020). *Deep Learning System to Screen Corona Virus Disease 2019 Pneumonia*. arXiv.

30. Zheng, C., Deng, X., et al., (2020). *Deep Learning-Based Detection for COVID-19 from Chest CT using Weak Label*. medRxiv.

31. Chen, J., Wu, L., et al., (2020). *Deep Learning-Based Model for Detecting 2019 Novel Coronavirus Pneumonia on High-Resolution Computed Tomography: A Prospective Study*. medRxiv.

32. Zhou, Z., Siddiquee, M. M. R., Tajbakhsh, N., & Liang, J., (2019). *Unet++: A Nested U-Net Architecture for Medical Image Segmentation.* IEEE TMI.
33. Song, Y., Zheng, S., et al., (2020). *Deep Learning Enables Accurate Diagnosis of Novel Coronavirus (COVID-19) with CT Images.* medRxiv.
34. Ronneberger, O., Fischer, P., & Brox, T., (2015). *U-Net: Convolutional Networks for Biomedical Image Segmentation.* In MICCAI. Springer.
35. Gozes, O., Frid-Adar, M., et al., (2020). *Rapid AI Development Cycle for the Coronavirus (COVID-19) Pandemic: Initial Results for Automated Detection & Patient Monitoring Using Deep Learning CT Image Analysis.* arXiv.
36. Tang, Z., Zhao, W., et al., (2020). *Severity Assessment of Coronavirus Disease 2019 (COVID-19) Using Quantitative Features from Chest CT Images.* arXiv.
37. Shi, F., Xia, L., et al., (2020). *Large-Scale Screening of COVID-19 from Community-Acquired Pneumonia Using Infection Size-Aware Classification.* arXiv.
38. Rajinikanth, V., Dey, N., et al., (2020). *Harmony-Search and Otsu based System for Coronavirus Disease (COVID-19) Detection Using Lung CT Scan Images.* arXiv.
39. Chaganti, S., Balachandran, A., et al., (2020). *Quantification of Tomographic Patterns Associated with COVID-19 from Chest CT.* arXiv.
40. Shan, F., Gao, Y., et al., (2020). *Lung Infection Quantification of COVID-19 in CT Images with Deep Learning.* arXiv.
41. Christe, A., et al., (2019). Computer-aided diagnosis of pulmonary fibrosis using deep learning and CT images. *Invest Radiol.* doi.org/10.1097/RLI.0000000000000574.
42. Bermejo-Pela´ez, D., Ash, S. Y., Washko, G. R., San, J. E. R., & Ledesma-Carbayo, M. J., (2020). Classification of interstitial lung abnormality patterns with an ensemble of deep convolutional neural networks. *Sci Rep.* doi.org/10.1038/s41598-019-56989-5.
43. Zhang, K., et al., (2020). Clinically applicable ai system for accurate diagnosis, quantitative measurements, and prognosis of COVID-19 pneumonia using computed tomography. *Cell, 181.* doi.org/10.1016/j.cell.2020.04.045.
44. Ronneberger, O., Fischer, P., & Brox, T., (2015). U-net: Convolutional networks for biomedical image segmentation. *Proceedings of the International Conference on Medical Image Computing and Computer-Assisted Intervention.*
45. Devalla, S. K., Renukanand, P. K., Sreedhar, B. K., Subramanian, G., Zhang, L., Perera, S., Mari, J. M., Chin, K. S., Tun, T. A., Strouthidis, N. G., et al., (2018). DRUNET: A dilated-residual U-net deep learning network to segment optic nerve head tissues in optical coherence tomography images. *Biomed. Opt. Express, 9.*
46. Long, J., Shelhamer, E., & Darrell, T., (2015). Fully convolutional networks for semantic segmentation. *Proceedings of the IEEE Conference on Computer Vision and Pattern Recognition.*
47. Badrinarayanan, V., Kendall, A., & Cipolla, R., (2017). SegNet: A deep convolutional encoder-decoder architecture for image segmentation. *IEEE Trans. Pattern Anal. Mach. Intell., 39.*
48. Chen, L. C., Papandreou, G., Schroff, F., & Adam, H., (2017). *Rethinking Atrous Convolution for Semantic Image Segmentation.* arXiv.
49. Yan, T., Wong, P. K., Ren, H., Wang, H., Wang, J., & Li, Y., (2020). Automatic distinction between COVID-19 and common pneumonia using multi-scale convolutional neural network on chest CT scans. *Chaos Solitons Fractals.* doi: 10.1016/j.chaos.2020.110153.

50. Ouyang, X., et al., (2020). Dual-sampling attention network for diagnosis of COVID-19 from community-acquired pneumonia. In: *IEEE Transactions on Medical Imaging.* doi: 10.1109/TMI.2020.2995508.

51. Bai, H. X., Hsieh, B., Xiong, Z., Halsey, K., et al., (2020). Performance of radiologists in differentiating COVID-19 from non-COVID-19 viral pneumonia at chest CT. *Radiology.* doi: 10.1148/radiol.2020200823.

52. Xu, X., Jiang, X., & Ma, C., (2020). A Deep learning system to screen novel coronavirus disease 2019 pneumonia. *Engineering.* doi.org/10.1016/j.eng.2020.04.010.

53. Kang, H., et al., (2020). Diagnosis of coronavirus disease 2019 (COVID-19) with structured latent multi-view representation learning. *IEEE Transactions on Medical Imaging.*

54. Khagi, B., Kwon, G. R., & Lama, R., (2019). Comparative analysis of Alzheimer's disease classification by CDR level using CNN, feature selection, and machine-learning techniques. *Int. J. Imaging Syst. Technol.*

55. Seetha, J., & Raja, S. S., (2018). Brain tumor classification using convolutional neural networks. *Biomedical & Pharmacology Journal.*

56. Alex Krizhevsky, Ilya Sutskever, & Geoffrey E. Hinton (2012). ImageNet classification with deep convolutional neural networks. In *Proceedings of the 25th International Conference on Neural Information Processing Systems – Volume 1 (NIPS'12).* Curran Associates Inc., Red Hook, NY, USA, 1097–1105.

57. Liu, S., & Deng, W., (2015). Very deep convolutional neural network-based image classification using small training sample size. In: 2015 *3rd IAPR Asian Conference on Pattern Recognition (ACPR).* doi.org/10.1109/acpr.2015.7486599.

58. Szegedy, C., et al., (2015). Going deeper with convolutions. In: *2015 IEEE conference on computer vision and pattern recognition (CVPR).* doi.org/10.1109/CVPR.2015.7298594.

59. He, K., Zhang, X., Ren, S., & Sun, J., (2016). Deep residual learning for image recognition. In: *2016 IEEE Conference on Computer Vision and Pattern Recognition (CVPR).* doi.org/10.1109/CVPR.2016.90.

60. Rahimzadeh, M., & Attar, A., (2020). *A New Modified Deep Convolutional Neural Network for Detecting COVID-19 from X-ray Images.* arxiv.

61. Alqudah, A. M., Qazan, S., Alquran, H. H., Alquran, H., Qasmieh, I., & Alqudah, A., (2020). A. *COVID-2019 Detection Using X-ray Images and Artificial Intelligence Hybrid Systems.* doi.org/10.13140/RG.2.2.16077.59362/1.

62. Fan, G., Chen, F., Chen, D., & Dong, Y., (2020). Recognizing multiple types of rocks quickly and accurately based on lightweight CNNs model. *IEEE Access.* doi. org/10.1109/ACCESS.2020.2982017.

63. Bandyopadhyay, S. K., & Dutta, S., (2020). *Machine Learning Approach for Confirmation of COVID-19 Cases: Positive, Negative.* MedRxiv: Death and Release. doi.org/10.1101/2020.03.25.20043505.2020.03.25.20043505.

64. Chen, Y., Gou, X., et al., (2019). Bone suppression of chest radiographs with cascaded convolutional networks in wavelet domain. *IEEE Access.* doi.org/10.1109/ACCESS.2018.2890300.

65. Kumar, P., & Kumari, S. (2020). *Detection of Coronavirus Disease (COVID-19) Based on Deep Features.* https://doi.org/10.33889/IJMEMS.2020.5.4.052 (accessed on 14 May 2022).

66. Horry, M. J., Chakraborty, S., Paul, M., Ulhaq, A., & Pradhan, B., (2020). *X-Ray Image-Based COVID-19 Detection Using Pre-Trained Deep Learning Models.* In Paul, M. (2020). COVID-19 Detection. Retrieved from osf.io/2uqdg.

67. Hemdan, E. E. D., Shouman, M. A., & Karar, M. E., (2020). *COVIDX-Net: A Framework of deep Learning Classifiers to Diagnose COVID-19 in X-ray Images.* arxiv.

68. Brunese, L., et al., (2020). Explainable deep learning for pulmonary disease and coronavirus COVID-19 detection from X-rays. *Computer Methods and Programs in Biomedicine.* doi.org/10.1016/j.cmpb.2020.105608.

69. Ucar, F., & Korkmaz, D., (2020). COVIDiagnosis-Net: Deep Bayes-SqueezeNet based diagnosis of the coronavirus disease 2019 (COVID-19) from X-ray images. *Medical Hypotheses.* doi.org/10.1016/j.mehy.2020.109761.

70. Wang, L., et al., (2020). *COVID-Net: A Tailored Deep Convolutional Neural Network Design for Detection of COVID-19 Cases from Chest X-ray Images.* arXiv.

71. Nour, M., Comert, Z., & Polat, K., (2020). A novel medical diagnosis model for COVID-19 infection detection based on deep features and Bayesian Optimization. *Applied Soft Computing Journal.* doi.org/10.1016/j.asoc.2020.106580.

72. Das, N. N., Kumar, N., et al., (2020). Automated deep transfer learning-based approach for detection of COVID-19 infection in chest X-rays. *IRBM.* doi.org/10.1016/j.irbm.2020.07.001.

73. Ozturk, T., et al., (2020). Automated detection of COVID-19 cases using deep neural networks with X-ray images. *Computers in Biology and Medicine.* doi.org/10.1016/j.compbiomed.2020.103792.

74. Oh, Y., Park, S., & Ye, J. C., (2020). Deep learning COVID-19 features on CXR using limited training data sets. *IEEE Transactions on Medical Imaging.* 10.1109/TMI.2020.2993291.

75. Rajaraman, R., et al., (2020). *Iteratively Pruned Deep Learning Ensembles for COVID-19 Detection in Chest.* arXiv.

76. Harris, S. L., & Harris, D. M., (2016). 3-sequential logic design. In: Harris, S. L., & Harris, D. M., (eds.), *Digital Design and Computer Architecture* (pp. 108–171). Boston: Morgan Kaufmann.

77. Fong, S. J., Li, G., Dey, N., Crespo, R. G., & Herrera-Viedma, E., (2020). Composite Monte Carlo decision making under high uncertainty of novel coronavirus epidemic using hybridized deep learning and fuzzy rule induction. *Appl Soft Comput J.* doi.org/10.1016/j.asoc.2020.106282.

78. Zhao, J., Zhang, Y., He, X., & Xie, P., (2020). *COVID-CT-Dataset: A CT Scan Dataset About COVID-19.* arXiv.

79. Cohen, J. P., Morrison, P., & Dao, L., (2020). *COVID-19 Image Data Collection.* arXiv.

CHAPTER 11

Generative Model and Its Application in Brain Tumor Segmentation

AMIT VERMA

School of Computer Science, UPES, Dehradun, Uttarakhand, India

ABSTRACT

A brain tumor is common cancer, nowadays, which can be categorized in majorly two classes that are malignant and benign. Benign is stable cancer that grows at a very negligible rate, but a malignant tumor grows exponentially. For the identification of brain tumor, the most appropriate technique is MR imaging, in which, with magnetic resonance the images of the brain is acquired. The operator, based on his experience, builds a report based on MR images of the patient. This report becomes the base of the diagnosis by the doctor. As the analysis of MR images is a tedious task which also lacks uniformity in the report as per the experience of MRI machine operator in various centers. It becomes the greater need of a medical field to automate the procedure of brain tumor segmentation with accuracy for a better understanding of the tumor size by the doctor. To solve the purpose, various techniques based on machine and DL play a vital role, but as the healthy tissues are always higher than the malignant tissues the trained model remains biased for positive tissues. Therefore, the generative model is used to generate the fair dataset to train the model which can give the results with higher accuracy and solve the problem of overfitting. In this chapter, the basic working of the generative model is discussed with the autoencoder (AE) approach and its application in segmenting the brain tumor.

Application of Deep Learning Methods in Healthcare and Medical Science.
Rohit Tanwar, PhD, Prashant Kumar, PhD, Malay Kumar, PhD, & Neha Nandal, PhD (Editors)
© 2023 Apple Academic Press, Inc. Co-published with CRC Press (Taylor & Francis)

11.1 INTRODUCTION TO GENERATIVE MODEL

Generative modeling [1] is used for estimating the density of the data, which is a problem in unsupervised learning. Unsupervised data [2] is data without a label that is 'x' is the data with no label. For unsupervised learning or to train a model based on unsupervised data, an underlying structure in the data is analyzed to propose a hypothetical model, which could be done using some clustering approach. The main objective of generative modeling is to train a model that represents the distribution of the data in the provided training data. Generative modeling is broadly used in the density estimation and generative new sample data from the given data. The generative approach uncovers the over and underlying features in the data set to create a more unbiased data set for training the classification model [3–5]. The main objective of the generative model is to find the way of learning these latent variables with the given observed data [6]. One of the simple generative models is AEs, which feed raw high dimensional data 'D' into low dimensional latent space 'Z,' where 'Z' can be thought of as a compressed representation of data 'D.' As in the case of clinical images, the dataset is a high dimensional dataset so, the encoder model can be applied to convert the data into low dimensional compressed data for better analysis and processing. Now we can use this compressed model 'Z' to generate a dataset resemble 'D' to train the model.

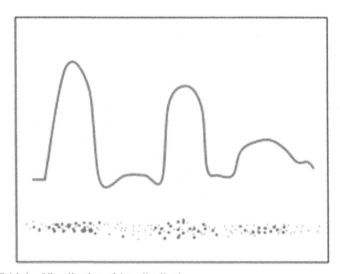

FIGURE 11.1   Visualization of data distribution.

11.1.1   DENSITY ESTIMATION

It is defined as the order of distribution of data in the provided samples. Having a bunch of samples that lie along with some probability distribution and we want to learn a model that approximates the distribution of the data as shown in Figure 11.1. Figure 11.1 is the visualization of the distribution of the data.

It is important to know the probability distribution of the data to choose the appropriate learning model or to know the anomaly in the data which is to be removed before using the data to train the model.

11.1.2   SAMPLE GENERATION

Another task of generative modeling is sample generation, idea is to generate a new sample from the input dataset 'd' or original data, representing 'd.' Example image-to-image conversion like having an image of the day can be used to get the same image of the night as shown in Figure 11.2. Figure 11.2(a) is an original image or data and from that data, a fake image of the evening sky is generated using the original data. Our model can be biased with the features that are overrepresented in the dataset without even knowing it. Here generative model can be used to create a fair dataset to train the model for unbiased results.

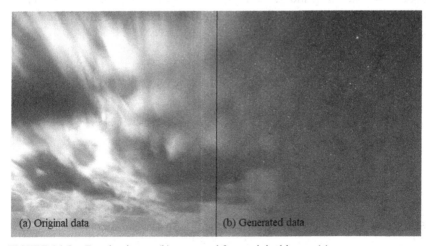

(a) Original data   (b) Generated data

FIGURE 11.2   Evening image (b) generated from original image (a).

11.2 BRAIN TUMOR SEGMENTATION USING AUTOENCODER APPROACH

Nowadays, segmentation of malignant parts in the brain automatically is one of the emerging research topics where DL is playing a vital role. As accurate detection of the tumor part in the brain is very important for a doctor to provide the best treatment to the patient [6–8]. The manual segmentation of the malignant tissues by the radiologist based on his or her experience and using some graphical tools is a tedious task. So, it is a great requirement of a medical field to have some automatic way to detect the tumor part in the brain without and human intervention with higher accuracy. Since tumor can be of variable shape and size, covering a very small portion of an MR image, it is a very challenging task to design a model which can almost accurately predict the boundary of the tumor with in the MRI. Many approaches based on machine learning were proposed to solve the purpose, recently researchers are using DL methods to train a model for automatic detection of malignant parts of the brain. The generative model-based AE approach is also a part of DL which can be used to train a model for autodetection of malignant tissue area among the healthy tissues. Autoencoding means automatic encoding data [9], it comprises of encoder and decoder both, encoder compress the high dimensional data 'x' into low dimensional data '$\bar{x}$' and decoder tries to extract the data say 'φ' from '$\bar{x}$' such the difference between x and '$\bar{x}$' is minimum. The process of encoding and decoding is represented in Figures 11.3 and 11.4, respectively. And the loss 'l' that is the difference between 'x' and 'φ' is shown in Figure 11.5.

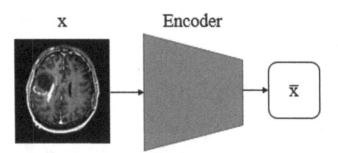

FIGURE 11.3 Encoding the original high dimensional data into low dimensional data.

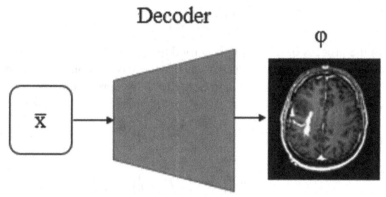

FIGURE 11.4   Decoding the data from low dimensional compressed dataset.

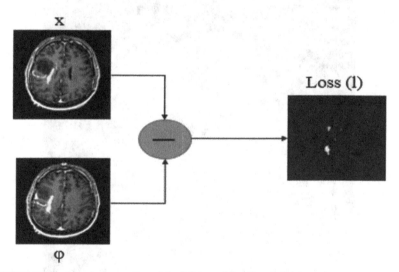

FIGURE 11.5   Loss between the original data and the decoded data.

The main objective of autoencoding is to minimize error. Let the data set be $D(x)$ as we can see that the data is having only one variable that is x without the output set 'y' as the data is unsupervised data. The encoder function 'f' is applied on data set $D(x)$ to get the low dimensional and compressed data set $D(\bar{x})$ as shown in Eqn. (1):

$$D(\bar{x}) = f\left(D(\bar{x})\right) \tag{1}$$

In the process of decoding, it is tried to access data set, let it be $D(\varphi)$ from compressed data set in such a way that the loss 'L' should be minimum as shown in Eqn. (3). As shown in Eqn. (2), 'β' is representing the decoder function applied on the dataset $D(\bar{x})$ to get the new dataset say $D(\varphi)$.

$$D(\varphi) = \beta\left(D(\bar{x})\right) \tag{2}$$

$$L = \left(\left\|D(\ ) - D(\bar{x})\right\|\right)^2 \tag{3}$$

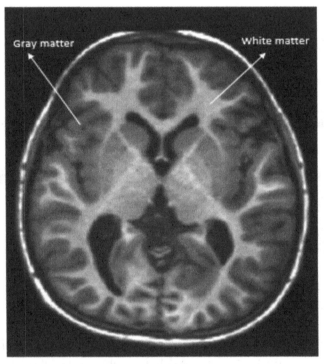

FIGURE 11.6   Representing white and gray matter in the MR image of a healthy brain.

Generative models use the spatial information of each tissue of the brain based on an atlas. Where brain tumor atlas is a common space having images of the brain with white matter (WM), gray matter (GM), and cerebrospinal fluid (CSF). GM and WM [10] are used to divide the complex anatomical structure of the brain, where GM is contained most of the neurons somas [11] and WM represents myelinated axons [12] as shown in Figure 11.6.

CSF [13] is a colorless liquid that surrounds the brain and spinal cord to protect it from any external or chemical damage. Spatial VAE [14, 15] that is spatial variational AE is one of the major states of art method. It became a popular method that allows making a probabilistic model using the latent variable. Pixel by pixel loss L is measured in the reconstructed images and the original image. VAE is a solution to the problem that is faced in the normal AE, when the AE creates a latent space that is the compressed form of the original data. The latent space is created in a discrete form that is data is classified in various classes. And the problem arises with the interpolation of the discrete data to generate new data. VAE allows us to have a continuous latent space instead of discrete that provides a powerful way to generate a new data set [16]. It generates two vectors [17] instead of a single vector of size n, which are mean vector (μ) and standard deviation vector (σ) as shown in Figure 11.7.

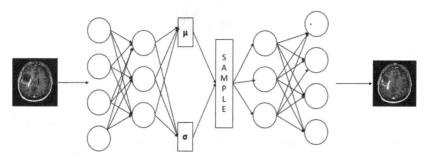

FIGURE 11.7 VAE model for generation brain image with malignant tissues.

In this case, the mean (μ) and standard deviation vector (σ) are calculated from the data to generate a continuous latent space. VAEs are applied to dense NNs and single-dimensional data [18, 19], which could be one of the disadvantages of VAE. And the main advantage of the AEs that AEs having the ability to generate sharper images [20]. It has been shown that the combination of VAE and AEs can produce better results. By combining the VAE and generative adversarial networks (GAN) to produce better results, in which the decoder part of AE is combined with the GAN generator. GAN uses both the images that is one of which is generated as a decoded image using AE and another one is the original image to generate a better result as shown in Figure 11.8. In AE the original image dataset represented by x is encoded to low dimensional or compressed data set z and from the compressed data set, a new dataset x' is constructed. Now the decoding part of AE is taken as a generator for GAN and compared with the original image x for the error.

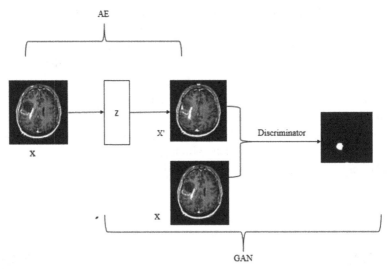

FIGURE 11.8 Combination of VAE and GAN to produce better results.

In the α-GAN [21] model, a hybrid approach is proposed by combining AE and GANs. This approach overcomes the drawbacks of both the VAE and GANs model, VAE often produces the blur images decoded from the compressed dataset z, whereas GANs suffer from the problem of mode collapsed. GAN model has a major challenge of evaluation which is overcome by the fusion of both the models. There are multiple applications of GAN as listed in Table 11.1. GANs is having a wide scope in many applications, it can be used to generate a fair dataset related to any problem. There are multiple fields in which we are having a limited datasets or a biased dataset therefore it become very difficult to train a model in efficient way. With GAN a fair dataset can be generated to train a machine in a much efficient way. This approach can be used for predicting the brain tumor using MR images for generating the fair dataset. As in the MR images of brain tumor most of the pixels represent the healthy tissue and very less pixels represent the malignant tissues, therefore it become difficult to train a fair model using the biased dataset. So, GAN can be used for generating the fair dataset on the basis of the biased dataset for more efficient results.

TABLE 11.1 Applications of GAN

Applications of GAN	
1. Brain tumor	Generating the fair (unbiased) dataset to train a model for better prediction.
2. Mixing images	Combining various image to generate a new image.
3. Plausible images	To generate plausible image dataset of handwritten characters to train the model more efficiently and accurately.
4. Human faces	To generate remarkably realistic pictures of humans.
5. Cartoon characters	Generating datasets of cartoon images to train a model.
6. Image translation	Converting images like from pencil sketch to color images, changing from one image to another, and many more.
7. Converting text to image	Taking input as text and generating different images out of it.

11.3 CONCLUSION AND FUTURE SCOPE

In this chapter, the basic generative model is discussed and its application in the segmentation of the brain tumor based on MR images. Generative models are an emerging approach in DL which has become a great interest for most researchers. Generative models are used to generate new images from the original datasets, it is helpful when we have a lack of data to train a model. Like in the case of MR images of brain tumors, we have a lot of data for healthy tissues but very little data regarding the malignant tissues. By the use of generative model data for malignant tissues for a brain tumor can be generated to train a model. This also resolves the problem of unbalanced data sets, which requires the preprocessing of data sets by applying to bag or boosting techniques. Further drawbacks of AEs and VAE are discussed and models representing the combination of AEs and VAE are also explained. GANs can have a wide application in the field of medical science for training the model with a fair dataset for higher accuracy. In the future, GAN can be used in various fields which remain untouched due to lack of proper and fair datasets, using GAN a fair dataset can be generated to train a model with more efficient and fair data.

KEYWORDS

- autoencoding
- gray matter
- hybrid approach
- neurons somas
- single-dimensional data
- white matter

REFERENCES

1. Wittrock, M. C., (1974). Learning as a generative process. *Educational Psychologist, 11*(2), 87–95.
2. Dash, M., Liu, H., & Yao, J., (1997). Dimensionality reduction of unsupervised data. In: *Proceedings Ninth IEEE International Conference on Tools with Artificial Intelligence* (pp. 532–539). IEEE.
3. Wittrock, M. C., (1974). A generative model of mathematics learning. *Journal for Research in Mathematics Education*, 181–196.
4. Zhong, S., & Ghosh, J., (2005). Generative model-based document clustering: A comparative study. *Knowledge and Information Systems, 8*(3), 374–384.
5. He, K., Wang, Y., & Hopcroft, J., (2016). *A Powerful Generative Model Using Random Weights for the Deep Image Representation.* arXiv preprint arXiv:1606.04801.
6. Chen, X., Mishra, N., Rohaninejad, M., & Abbeel, P., (2018). PixelSNAIL: An improved autoregressive generative model. In: *International Conference on Machine Learning* (pp. 864–872). PMLR.
7. Raja, P. S., (2020). Brain tumor classification using a hybrid deep autoencoder with Bayesian fuzzy clustering-based segmentation approach. *Biocybernetics and Biomedical Engineering, 40*(1), 440–453.
8. Siva, R. P. M., & Rani, A. V., (2020). Brain tumor classification using a hybrid deep autoencoder with Bayesian fuzzy clustering-based segmentation approach. *Biocybernetics and Biomedical Engineering, 40*(1).
9. Chen, L., Xie, Y., Sun, J., Balu, N., Mossa-Basha, M., Pimentel, K., & Yuan, C., (2017). 3D intracranial artery segmentation using a convolutional autoencoder. In: *2017 IEEE International Conference on Bioinformatics and Biomedicine (BIBM)* (pp. 714–717). IEEE.

10. Huang, H., Zhang, J., Wakana, S., Zhang, W., Ren, T., Richards, L. J., & Mori, S., (2006). White and gray matter development in human fetal, newborn, and pediatric brains. *Neuroimage, 33*(1), 27–38.

11. Kunze, W. A. A., Clerc, N., Furness, J. B., & Gola, M., (2000). The soma and neurites of primary afferent neurons in the guinea-pig intestine respond differentially to deformation. *The Journal of Physiology, 526*(2), 375–385.

12. Poliak, S., & Peles, E., (2003). The local differentiation of myelinated axons at nodes of Ranvier. *Nature Reviews Neuroscience, 4*(12), 968–980.

13. Seehusen, D. A., Reeves, M., & Fomin, D., (2003). Cerebrospinal fluid analysis. *American Family Physician, 68*(6), 1103–1108.

14. Rezende, D. J., Mohamed, S., & Wierstra, D., (2014). Stochastic backpropagation and approximate inference in deep generative models. In: *International Conference on Machine Learning* (pp. 1278–1286). PMLR.

15. Kingma, D. P., & Welling, M., (2013). *Auto-Encoding Variational Bayes.* arXiv preprint arXiv:1312.6114.

16. Tomczak, J., & Welling, M., (2018). VAE with a VampPrior. In: *International Conference on Artificial Intelligence and Statistics* (pp. 1214–1223). PMLR.

17. Burgess, C. P., Higgins, I., Pal, A., Matthey, L., Watters, N., Desjardins, G., & Lerchner, A., (2018). *Understanding Disentangling in $\Beta $-VAE.* arXiv preprint arXiv:1804.03599.

18. An, J., & Cho, S., (2015). Variational autoencoder based anomaly detection using reconstruction probability. *Special Lecture on IE, 2*(1), 1–18.

19. Kingma, D. P., & Welling, M., (2013). *Auto-Encoding Variational Bayes.* arXiv preprint arXiv:1312.6114.

20. Baur, C., Wiestler, B., Albarqouni, S., & Navab, N., (2018). Deep autoencoding models for unsupervised anomaly segmentation in brain MR images. In: *International MICCAI Brain Lesion Workshop* (pp. 161–169). Springer, Cham.

21. Rosca, M., Lakshminarayanan, B., Warde-Farley, D., & Mohamed, S., (2017). *Variational Approaches for Auto-encoding Generative Adversarial Networks.* arXiv preprint arXiv:1706.04987.

CHAPTER 12

Genomic Sequence Similarity of SARS-CoV2 Nucleotide Sequences Using Biopython: Key for Finding Cure and Vaccines

SWEETI SAH,[1] B. SURENDIRAN,[1] and R. DHANALAKSHMI[2]

[1]*Department of Computer Science and Engineering, National Institute of Technology Puducherry, Karaikal, India, E-mails: sweetisah3@gmail.com (S. Sah), surendiran@nitpy.ac.in (B. Surendiran)*

[2]*IIIT Tiruchirappalli, Trichy, Tamil Nadu, India, E-mail: r_dhanalakshmi@yahoo.com (R. Dhanalakshmi)*

ABSTRACT

Genomic analyzes of coronaviruses are crucial as understanding the properties of genomic sequences is still a challenging task. Genomic sequence analysis consists of finding similarities between sequences, finding inherent features of the sequences, finding sequence variations, its evolution, genetic diversity, and also molecular structure. In order to analyze the genomic sequences, the similarity between the nucleotide sequences of SARS-CoV2 from different countries has been analyzed using the dot plot graph method. Based on the dataset, collected from National Center for Biotechnology Information website, computation of guanine-cytosine (GC) content percentage is done from the given sequences and graphically visualized the nucleotide ratio for a particular sequence. Moreover, a dot plot graph

Application of Deep Learning Methods in Healthcare and Medical Science.
Rohit Tanwar, PhD, Prashant Kumar, PhD, Malay Kumar, PhD, & Neha Nandal, PhD (Editors)

is constructed to find the similarities between nucleotides by plotting a dot in the matrix using biopython library. The result showed that the sequences are homologous to each other. The nucleotide sequences consist of all the information encoded in the virus. Hence, understanding genetic information plays a vital role in finding cures and vaccines for the disease.

12.1   INTRODUCTION

COVID-19 or SARS-CoV2 was first raised in Wuhan, China which brought a pandemic and dreadful situation all across the world [12]. Genomic analysis is required to understand genetic variability [11]. Vaccine development is time absorbing process and needs analysis for genetic variability, so vaccines can be made for the human population [13]. As protein is a crucial element of the living organism, it is responsible for several activities such as metabolism, DNA replication, hormone regulation, and transcription [6, 7]. Protein is a chain of amino acid and each protein has a unique amino acid sequence [8]. Figure 12.1 shows the SARS-CoV2 genomic orientation [11].

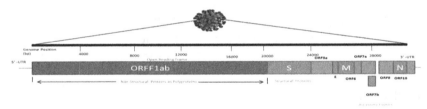

FIGURE 12.1   SARS-CoV2 genomic orientation.
Source: Adapted from Ref. [11].

The metagenomic analysis is done through next-generation sequencing [15] explains that SARS-CoV2 is single-stranded enveloped. Also, RNA virus with 29.9 kilobases of genome length [16–19]. There are a total of 11 coding regions as per National Center for Biotechnology Information Web Source [10] that cipher ORF1ab polyproteins, envelope (E) protein, Membrane (M) glycoprotein, nucleocapsid (N) protein, spike (S) glycoprotein, and also other proteins like ORF3a protein, ORF6 protein, ORF7a protein, ORF7b protein, ORF8 protein as well as ORF10 protein whereas nonstructural proteins (NSP) ciphered from an open reading frame (ORF).

Amino acids are made up of a set of 20 different small molecules. Protein consists of an amino acid chain or polypeptides. This sequence only causes the polypeptide to fold into a shape. Amino acid sequences are encoded in genes. Alignment of sequences is done to classify protein families, recognition of parallel transferred genes as well as the sequence of recombined sequences (Figure 12.2) [20].

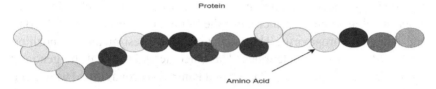

FIGURE 12.2   Protein consisting sequence of amino acids.

There are several alignment tools like BLAST [21], FASTA [22] (similarity search tools) and muscle [24], ClustalW [23], MAFT [25] multiple sequence alignment (MSA). For Sequences' profile search programs tools such as PSI-BLAST [21], HMMER/pfam [26] and for genome alignment tools such as progressive Mauve [27], BLASTZ [28], TBA [29]. These tools help us to assess the functions of genes as well as proteins.

The chapter is organized as follows; Section 12.1 consists of an introduction. Section 12.2 shows the literature survey of work done for sequence, identification, and prediction. Section 12.3 shows the proposed methodology. Sections 12.4–12.6 show the experimental analysis of the genomic dataset using biopython; dataset description; and experimental results, respectively. Finally, Section 12.7 shows the result and conclusion produced after finding the percentage similarity between nucleotide sequences.

12.2 LITERATURE REVIEW

Similar proteins sequence identification is the main step in computational biology pipelines like identifying similar sequences of protein, also the origination of more protein like graphs for functional annotation, downstream analysis as well as for the location of the gene [1]. This chapter proposed a work for distributed memory software called PASTIS which relies on sparse matrix calculation for the better recognition of similar proteins. Sparse matrix was used for scalability and also it fits well for protein sequence synonymity search when binding with completely distributed dictionary comprising genomic sequences, which allowed sequences requests to be granted that are remote. The proposed algorithm does not alter the basic sparse matrix model and also attains good scaling up to millions of protein sequences.

Touati et al. [4] examined the SARS-CoV-2 genomic signature through various nucleotide representations as well as signal processing tools to recognize the generic beginning of viruses. It was then correlated with 21 other suitable sequences including the genomic sequences of Yak, Bat, and Pangolin coronavirus. They developed the latest method to position the modifications of nucleotides. The outcome showed that the genomic sequences of coronaviruses like Bat and Pangolin coronaviruses were quite similar to SARS-CoV-2 sequences with a similarity percentage of 96% and 86% all along the genome. Pangolin coronavirus sequence showed the local highest nucleotide identity within the S gene sequence. The overall study finds a way to automatically characterize the viruses through their sequences and gives an opportunity for quick classification and recognition of virus origin. Thus, their technique can extract the numerical features to classify the several coronaviruses.

Kulmanov and Robert [5] developed a method to predict functions of protein from sequences that join deep CNN with genomic sequence resemblance-based forecasting. The model examines the genomic sequences that can forecast the function of protein and then joins with the purpose of the same proteins. The accomplishment of DeepGOPlus with the help of CAFA3 evaluation measure achieves an F_{max} of 0.390, 0.614, and 0.557 for BPO, CCO, and MFO respectively. CCO showed the best prediction as compared to BPO and MFO. Also compared DeepGOPlus with other methods like GOLabeler and DeepText2GO on another dataset as well. DeepGOPlus can explain approx. 40 sequences of protein per sec on usual hardware, hence making quick and also, more precise purpose for protein function on broad range.

Ceraolo and Federico [9] analyzed 56 genomics sequences from different 2019-nCoV patients and showed the highest similarity sequence of 99%. The genomic sequence that appears close to 2019-nCoV are coronaviruses infecting through bats, while MERS and SARS are related distantly. Hence constructed phylogenetic tree and despite of low heterogeneity of genome it is able to identify at least two hypervariable genomic hotspots. Finally, identified the amino acid differences for antiviral strategies which is derived from past anti-coronavirus approaches.

Saha et al. [14] analyzed SARS-CoV-2, 566 Indian sequences with the help of Multiple Sequence Alignment (MSA) Techniques viz. MUSCLE, ClustalO, ClustalW, and MAFFT in order to line up and also to recognize record of alterations as insertion, substitution, deletion, and SNP. The consensus outcome Consensus Multiple Sequence Alignment (CMSA) was done to show the records of alterations. The results showed 7,672,025 and 54 unique substitutions, SNPs, and deletions in SARS-CoV-2 Indian genomes and 4 SNPs, 54 SNPs were near to 60% of the virus population. The outcome can be beneficial for virus categorization, defining, and designing the vaccine dose for the human population in India.

Zielezinski et al. [20] analyzed the number of total 566 genomes of Indian SARS-CoV-2 to show hereditary alteration as substitution, insertion, deletion, and SNP. The database is gathered from GISAID. 57 SNPs out of 64 SNPs discovered in six coding regions of SARS-CoV-2 genomes. Also, to show variations in virus, phylogenetic analysis was done. Hence, performed MSA for the sequences found from NCBI web source. After the alignment process, a consensus sequence was made and discovered 933 substitutions, 2 insertions, and 2,449 deletions in overall 3,384 exclusive alteration points, 566 genomic sequences beyond 29.9 Kbp. Classified 100 clusters of mutations, 1,609-point mutations, and 64 SNPs. The output is viewed using BioCircos, bar plots, and plotting entropy. SNPs are a beneficial target for the categorization of viruses, defining, and designing the effective dose of the COVID-19 vaccine.

Chang et al. [33] analyzed 10 sequences of SARS-CoV2 from National Center for Biotechnology Information for alignment of the genome to observed negligible dissimilarities in protein sequences with M protein and N protein. There were two variances in the Spike protein region. One alteration detected from the sequence of South Korea was confirmed. The other two possible "L" and "S" SNPs were detected from regions like ORF1ab and ORF8. Hence, performed the genomic analysis of sequences and compared or correlated multiple sequences of SARS-CoV2.

12.3 PROPOSED METHODOLOGY

Proposed methodology for analyzing the SARS-CoV2 nucleotide sequences is mentioned below (Figure 12.3):

- Load Fasta File consisting of Nucleotide Sequences of SARS-Cov2 viruses;
- Count the number of nucleotides for all the sequences;
- Calculate the guanine-cytosine (GC) content percentage;
- Find the similarity between the nucleotide sequence using a dot plot graph.

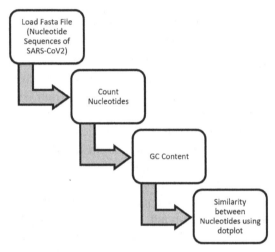

FIGURE 12.3   Workflow for analyzing the nucleotide of SARS-Cov2.

12.4   EXPERIMENTAL SETUP

The experimental setup requirements are listed in Table 12.1.

TABLE 12.1   Experimental Requirement Details

Libraries Used	Biopython
	Bio.SeqUtils.ProtParam
	Protein analysis
	matplotlib.pyplot
	NumPy
Dataset Download	NCBI website [10]
Platform	Colab
Language	Python

12.5   DATASET DESCRIPTION

The database used consists of a unique accession number and is collected from the NCBI website. The SARS-CoV2 dataset has FASTA format consisting of ID, Name, and Descriptions of coronaviruses [10] (Figure 12.4).

```
29903
ID: NC_045512.2
Name: NC_045512.2
Description: NC_045512.2 |Severe acute respiratory syndrome coronavirus 2 isolate Wuhan-Hu-1, complete genome
Number of features: 0
Seq('ATTAAAGGTTTATACCTTCCCAGGTAACAAACCAACCAACTTTCGATCTCTTGT...AAA')
29888
ID: LC581364.1
Name: LC581364.1
Description: LC581364.1 |Severe acute respiratory syndrome coronavirus 2 TKYE63319_2020 genomic RNA, complete genome
Number of features: 0
Seq('CTTCCCAGGTAACAAACCAACCAACTTTCGATCTCTTGTAGATCTGTTCTCTAA...AAA')
29900
ID: LC581365.1
Name: LC581365.1
Description: LC581365.1 |Severe acute respiratory syndrome coronavirus 2 TKYE63557_2020 genomic RNA, complete genome
Number of features: 0
Seq('AAAGGTTTATACCTTCCCAGGTAACAAACCAACCAACTTTCGATCTCTTGTAGA...AAA')
29897
ID: LC581366.1
Name: LC581366.1
Description: LC581366.1 |Severe acute respiratory syndrome coronavirus 2 TKYE63557d_2020 genomic RNA, complete genome
Number of features: 0
Seq('AAAGGTTTATACCTTCCCAGGTAACAAACCAACCAACTTTCGATCTCTTGTAGA...AAA')
29891
ID: LC581367.1
Name: LC581367.1
Description: LC581367.1 |Severe acute respiratory syndrome coronavirus 2 TKYE63558_2020 genomic RNA, complete genome
Number of features: 0
Seq('TACCTTCCCAGGTAACAAACCAACCAACTTTCGATCTCTTGTAGATCTGTTCTC...AAA')
29888
ID: LC581368.1
Name: LC581368.1
Description: LC581368.1 |Severe acute respiratory syndrome coronavirus 2 TKYE63558d_2020 genomic RNA, complete genome
Number of features: 0
Seq('TACCTTCCCAGGTAACAAACCAACCAACTTTCGATCTCTTGTAGATCTGTTCTC...AAA')
29891
```

FIGURE 12.4   SARS-CoV2 viruses dataset sample (Fasta format).

After reading the Fasta format file of several coronaviruses, the total count of nucleotide is computed (i.e., A, T, G, C) in order to analyze the particular sequences (Figure 12.5).

	Nucleotide	id	count
101	(A, C, T, T, T, C, G, A, T, C, T, C, T, T, G, ...	MW093533.1	29865
102	(C, T, T, T, C, G, A, T, C, T, C, T, T, G, T, ...	MW093534.1	29823
103	(C, T, T, T, C, G, A, T, C, T, C, T, T, G, T, ...	MW093535.1	29812
104	(C, T, T, T, C, G, A, T, C, T, C, T, T, G, T, ...	MW093536.1	29823
105	(A, C, T, T, T, C, G, A, T, C, T, C, T, T, G, ...	MW093537.1	503

FIGURE 12.5   Nucleotide count of several SARS-CoV2 viruses sequences.

Now, fetching the highest nucleotide count of several sequences, where id shows the unique identity of the virus.

12.6　EXPERIMENTAL RESULTS

Reading the sequence string from the FASTA file and then use the string count method to count the occurrences of nucleotides that how many times each letter has occurred. As the sequence consists of four alphabets that are Adenine (A), Cytosine (C), Guanine (G), Thymine (T) and are connected to each other through a chain-like structure by forming a covalent bond (Figures 12.6 and 12.7) [2].

	Nucleotide	id	count
0	(A, T, T, A, A, A, G, G, T, T, T, A, T, A, C, ...	NC_045512.2	29903
2	(A, A, A, G, G, T, T, T, A, T, A, C, C, T, T, ...	LC581365.1	29900
29	(A, A, G, G, T, T, T, A, T, A, C, C, T, T, C, ...	MW093459.1	29899
30	(A, A, G, G, T, T, T, A, T, A, C, C, T, T, C, ...	MW093460.1	29899
44	(A, A, G, G, T, T, T, A, T, A, C, C, T, T, C, ...	MW093474.1	29899
48	(A, A, G, G, T, T, T, A, T, A, C, C, T, T, C, ...	MW093478.1	29899
12	(A, G, G, T, T, T, A, T, A, C, C, T, T, C, C, ...	LC581375.1	29898
3	(A, A, A, G, G, T, T, T, A, T, A, C, C, T, T, ...	LC581366.1	29897
13	(A, G, G, T, T, T, A, T, A, C, C, T, T, C, C, ...	LC581376.1	29895
55	(T, T, A, T, A, C, C, T, T, C, C, C, A, G, G, ...	MW093485.1	29894
60	(T, A, T, A, C, C, T, T, C, C, C, A, G, G, T, ...	MW093490.1	29893
19	(A, T, A, C, C, T, T, C, C, C, A, G, G, T, A, ...	MW093449.1	29892
31	(A, T, A, C, C, T, T, C, C, C, A, G, G, T, A, ...	MW093461.1	29892
37	(A, T, A, C, C, T, T, C, C, C, A, G, G, T, A, ...	MW093467.1	29892
45	(T, T, T, A, T, A, C, C, T, T, C, C, C, A, G, ...	MW093475.1	29892
4	(T, A, C, C, T, T, C, C, C, A, G, G, T, A, A, ...	LC581367.1	29891
6	(T, A, C, C, T, T, C, C, C, A, G, G, T, A, A, ...	LC581369.1	29891
8	(T, A, C, C, T, T, C, C, C, A, G, G, T, A, A, ...	LC581371.1	29891
43	(T, A, C, C, T, T, C, C, C, A, G, G, T, A, A, ...	MW093473.1	29891
58	(T, A, C, C, T, T, C, C, C, A, G, G, T, A, A, ...	MW093488.1	29891

FIGURE 12.6　Nucleotide count of several SARS-CoV2 viruses sequences (Top 20).

Below, the GC content of SARS-Cov2 Sequences is computed using BioPython. In molecular biology or genetics, these are nitrogenous bases in DNA or RNA molecules. It is calculated as [31]:

$$((G+C)/(A+T+G+C))*100\% \qquad (1)$$

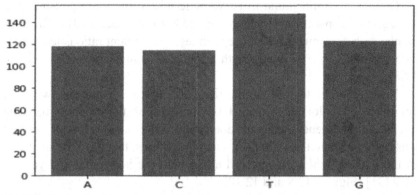

FIGURE 12.7   Graphical view of nucleotide count for particular sequence (sample).

```
GC content of covid_seq :: 37.97277865097148
GC content of covid_seq :: 37.97845289079229
GC content of covid_seq :: 37.97324414715719
GC content of covid_seq :: 37.973709736762885
GC content of covid_seq :: 37.97798668495534
GC content of covid_seq :: 37.97845289079229
GC content of covid_seq :: 37.98133217356395
GC content of covid_seq :: 37.981798715203425
GC content of covid_seq :: 37.97798668495534
GC content of covid_seq :: 37.97845289079229
GC content of covid_seq :: 37.97845289079229
GC content of covid_seq :: 37.97891919022921
GC content of covid_seq :: 37.98247374406315
GC content of covid_seq :: 37.982940291018565
GC content of covid_seq :: 37.25798011512297
GC content of covid_seq :: 37.20565149136578
GC content of covid_seq :: 37.10099424385139
GC content of covid_seq :: 37.307936933016435
GC content of covid_seq :: 37.662772403173435
GC content of covid_seq :: 37.642178509300145
GC content of covid_seq :: 37.43177632680395
GC content of covid_seq :: 37.63142034420411
GC content of covid_seq :: 37.64638961386603
GC content of covid_seq :: 37.8819351380327
GC content of covid_seq :: 37.93957238933316
GC content of covid_seq :: 36.963000167420056
GC content of covid_seq :: 37.632680395111336
GC content of covid_seq :: 35.5956606174245
GC content of covid_seq :: 37.753222836095766
GC content of covid_seq :: 37.64339944479749
GC content of covid_seq :: 37.663467005585474
GC content of covid_seq :: 37.63548775592132
GC content of covid_seq :: 36.68173447178972
GC content of covid_seq :: 37.04546976495011
GC content of covid_seq :: 37.17125075286087
GC content of covid_seq :: 37.63016037767436
GC content of covid_seq :: 37.595419847328245
GC content of covid_seq :: 37.80275659039208
GC content of covid_seq :: 37.086891009542946
```

FIGURE 12.8   GC content of SARS-CoV2 sequences (sample).

GC content ratio found to be variable with different organisms. The average GC content ratio ranges from 35% to 60% across 100-Kb fragments, with 41% mean. Sequence with less GC content ratio is less stable as compared to the sequence with a high GC content ratio [31] (Figure 12.8).

Now, in order to find similarities between two sequences, we have computed a dot plot matrix using biopython to show the similarity between nucleic acid sequences using the dot indicated if a match is found. Dot Plot was first introduced in 1970 [31]. The 2D matrices, that have the sequence of the nucleic acid are compared along with the horizontal and vertical axes [31] (Figures 12.9 and 12.10).

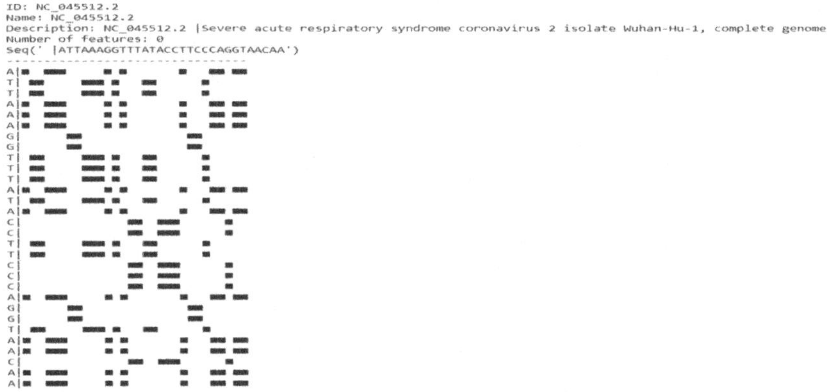

```
ID: NC_045512.2
Name: NC_045512.2
Description: NC_045512.2 |Severe acute respiratory syndrome coronavirus 2 isolate Wuhan-Hu-1, complete genome
Number of features: 0
Seq('  |ATTAAAGGTTTATACCTTCCCAGGTAACAA')
```

FIGURE 12.9 Dot plot between SARS-CoV2 sequences (sample).

12.7 CONCLUSION

Virus genomic structure is one of the factors that shows a great challenge to researchers, and also its ability to adapt to various climatic conditions. The proposed work consists of the analysis of the nucleotide sequence of coronavirus named SARS-CoV2 virus to find the graphical similarity between the sequences using a dot plot. The sequences have more than 30% similarity, so we can say the sequences are homologous, which means the sequence has common evolutionary ancestors also their structures and functions may have a common resemblance. This helps in analyzing the future sequences in a more visual way and helps in predicting the function of the unknown sequence.

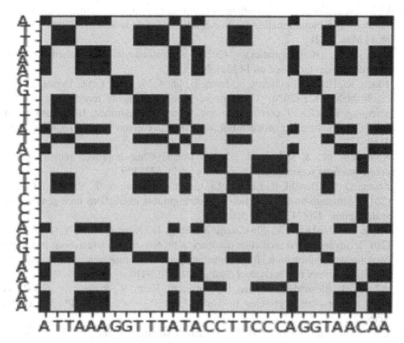

FIGURE 12.10   Colorful representation between SARS-CoV2 sequences using dot plot (sample).

KEYWORDS

- consensus multiple sequence alignment
- guanine-cytosine
- multiple sequence alignment
- nonstructural proteins
- open reading frame

REFERENCES

1. Selvitopi, O., Saliya, E., Giulia, G., Georgios, P., Ariful, A., & Aydin, B., (2020). *Distributed Many-to-Many Protein Sequence Alignment Using Sparse Matrices.* arXiv preprint arXiv:2009.14467.

2. Vijini, M., (2017). *Starting off in Bioinformatics.* https://towardsdatascience.com/starting-off-in-bioinformatics-dna-nucleotides-and-strands-8c32515271a8 (accessed on 14 May 2022).
3. Dot Plot (Bioinformatics), (2020). https://en.wikipedia.org/wiki/Dot_plot_ (bioinformatics) (accessed on 14 May 2022).
4. Touati, R., Haddad-Boubaker, S., Imen, F., Imen, M., Afef, E. O., Henda, T., Zied, L., & Maher, K., (2020). Comparative genomic signature representations of the emerging COVID-19 coronavirus and other coronaviruses: High identity and possible recombination between Bat and pangolin coronaviruses. *Genomics, 112*(6), 4189–4202.
5. Kulmanov, M., & Robert, H., (2020). DeepGOPlus: Improved protein function prediction from sequence. *Bioinformatics, 36*(2), 422–429.
6. Zhang, Q. C., Donald, P., Lei, D., Li, Q., Yu, S., Chan, A. T., Brygida, B., et al., (2012). Structure-based prediction of protein-protein interactions on a genome-wide scale. *Nature, 490*(7421), 556–560.
7. Wang, L., Zhu-Hong, Y., Shi-Xiong, X., Feng, L., Xing, C., Xin, Y., & Yong, Z., (2017). Advancing the prediction accuracy of protein-protein interactions by utilizing evolutionary information from position-specific scoring matrix and ensemble classifier. *Journal Of Theoretical Biology, 418*, 105–110.
8. Alberts, B., Alexander, J., Julian, L., Martin, R., Keith, R., & Peter, W., (2002). The shape and structure of proteins. In: *Molecular Biology of the Cell* (4th edn.). Garland Science.
9. Ceraolo, C., & Federico, M. G., (2020). Genomic variance of the 2019-nCoV coronavirus. *Journal of Medical Virology, 92*(5), 522–528.
10. National Center for Biotechnology Information (NCBI) [Internet], (1988). Bethesda (MD): National Library of Medicine (US), National Center for Biotechnology Information.
11. Saha, I., Nimisha, G., Debasree, M., Nikhil, S., & Kaushik, M., (2020). Inferring the genetic variability in Indian SARS-CoV-2 genomes using consensus of multiple sequence alignment techniques. *Infection, Genetics, and Evolution, 85*, 104522.
12. Zhu, N., Dingyu, Z., Wenling, W., Xingwang, L., Bo, Y., Jingdong, S., Xiang, Z., et al., (2020). A novel coronavirus from patients with pneumonia in China, 2019. *New England Journal of Medicine.*
13. Poland, G. A., (2020). Tortoises, hares, and vaccines: A cautionary note for SARS-CoV-2 vaccine development. *Vaccine, 38*(27), 4219.
14. Saha, I., Nimisha, G., Debasree, M., Nikhil, S., Jnanendra, P. S., & Kaushik, M., (2020). Genome-wide analysis of Indian SARS-CoV-2 genomes for the identification of genetic mutation and SNP. *Infection, Genetics, and Evolution, 85*, 104457.
15. Lu, I-Na, Claude, P. M., & Feng, Q. H., (2020). Applying next-generation sequencing to unravel the mutational landscape in viral quasispecies: A mini-review. *Virus Research*, 197963.
16. Cui, J., Fang, L., & Zheng-Li, S., (2019). Origin and evolution of pathogenic coronaviruses. *Nature Reviews Microbiology, 17*(3), 181–192.
17. Su, S., Gary, W., Weifeng, S., Jun, L., Alexander, C. K. L., Jiyong, Z., Wenjun, L., Yuhai, B., & George, F. G., (2016). Epidemiology, genetic recombination, and pathogenesis of coronaviruses. *Trends in Microbiology, 24*(6), 490–502.

18. Weiss, S. R., & Navas-Martin, S., (2005). Coronavirus pathogenesis and the emerging pathogen severe acute respiratory syndrome coronavirus. *Microbiology and Molecular Biology Reviews, 69*(4), 635–664.

19. Zhou, P., Xing-Lou, Y., Xian-Guang, W., Ben, H., Lei, Z., Wei, Z., Hao-Rui, S., et al., (2020). A pneumonia outbreak associated with a new coronavirus of probable bat origin. *Nature, 579*(7798), 270–273.

20. Zielezinski, A., Susana, V., Jonas, A., & Wojciech, M. K., (2017). Alignment-free sequence comparison: Benefits, applications, and tools. *Genome Biology, 18*(1), 186.

21. Altschul, S. F., Madden, T. L., Schäffer, A. A., Zhang, J., Zhang, Z., Miller, W., et al., (1997). Gapped BLAST and PSI-BLAST: A new generation of protein database search programs. *Nucleic Acids Res., 25*, 3389–3402.

22. Pearson, W. R., & Lipman, D. J., (1988). Improved tools for biological sequence comparison. *Proc. Natl. Acad. Sci U S A., 85*, 2444–2448.

23. Thompson, J. D., Higgins, D. G., & Gibson, T. J., (1994). CLUSTAL W: Improving the sensitivity of progressive multiple sequence alignment through sequence weighting, position-specific gap penalties and weight matrix choice. *Nucleic Acids Res., 22*, 4673–4680.

24. Edgar, R. C., (2004). MUSCLE: Multiple sequence alignment with high accuracy and high throughput. *Nucleic Acids Res., 32*, 1792–1797.

25. Katoh, K., Misawa, K., Kuma, K., & Miyata, T., (2002). MAFFT: A novel method for rapid multiple sequence alignment based on fast Fourier transform. *Nucleic Acids Res., 30*, 3059–3066.

26. Finn, R. D., Bateman, A., Clements, J., Coggill, P., Eberhardt, R. Y., Eddy, S. R., et al., (2014). Pfam: The protein families database. *Nucleic Acids Res., 42*, D222–230.

27. Darling, A. E., Mau, B., & Perna, N. T., (2010). Progressive mauve: Multiple genome alignment with gene gain, loss, and rearrangement. *PLoS One, 5*, e11147.

28. Schwartz, S., Kent, W. J., Smit, A., Zhang, Z., Baertsch, R., Hardison, R. C., et al., (2003). Human-mouse alignments with BLASTZ. *Genome Res., 13*, 103–107.

29. Blanchette, M., Kent, W. J., Riemer, C., Elnitski, L., Smit, A. F., Roskin, K. M., et al., (2004). Aligning multiple genomic sequences with the threaded blockset aligner. *Genome Res., 14*, 708–715.

30. Tai-Jay, C., De-Ming, Y., Mong-Lien, W., Kung-How, L., Ping-Hsing, T., Shih-Hwa, C., Ta-Hsien, L., & Chin-Tien, W., (2020). Genomic analysis and comparative multiple sequences of SARS-CoV2. *Journal of the Chinese Medical Association, 83*(6), 537–543.

31. GC-Content, (2021). https://en.wikipedia.org/wiki/GC-content (accessed on 14 May 2022).

32. Gibbs, A. J., & George, A. M., (1970). The diagram, a method for comparing sequences: Its use with amino acid and nucleotide sequences. *European Journal of Biochemistry, 16*(1), 1–11.

33. Chang, T. J., Yang, D. M., Wang, M. L., Liang, K. H., Tsai, P. H., Chiou, S. H., & Wang, C. T. (2020). Genomic analysis and comparative multiple sequences of SARS-CoV2. *Journal of the Chinese Medical Association, 83*(6), 537–543.

CHAPTER 13

Autonomous Logistic Transportation System for Smart Healthcare System

SASWATI KUMARI BEHERA[1] and G. PRASHANTH[2]

[1]*Department of Electrical and Electronics Engineering, Sri Sairam Engineering College, Chennai, Tamil Nadu, India, E-mail: saswatikumari.eee@sairam.edu.in*

[2]*Software Engineer, Wipro Technology, Bangalore, Karnataka, India*

ABSTRACT

At present, during this COVID-19 pandemic, where social distancing and the safety of health workers are most important, we can use autonomous logistic transportation system (ALTS) to provide medicine, and food to the COVID-19 patients. This ALTS will act as a crown in the smart healthcare system as far as the current situation is concerned. Transportation without a driver was developed as early as the 1950s for in-house logistics. Those robots were developed primarily for dangerous or practically inaccessible situations. In order to have automated cargo transportation, we employ computer vision-based techniques. The operations of complex traffic conditions of connected automated vehicles and the perception of the road environment can be developed, due to the improvement in the accuracy and functions of the controller. Since transportation adds no value to the product, we suggest a control strategy based on environment detection to save money while achieving the same impact as costly multiline radar. To have an optimistic approach for detecting lanes, traffic signs in the road we use computer vision-based functions. To achieve the required destination,

Application of Deep Learning Methods in Healthcare and Medical Science.
Rohit Tanwar, PhD, Prashant Kumar, PhD, Malay Kumar, PhD, & Neha Nandal, PhD (Editors)

we use GPS for transporting Cargo. Haar Cascade Classifier is used to detect and train objects.

13.1 INTRODUCTION

Autonomous technology is a common research area in computer science in which artificial intelligence (AI) and machine learning (ML) are only a few of the subjects covered. In recent years, DL (DL) also displayed its enormous processing capacity by expanding its capabilities into increasingly complex areas.

DL is being used as a means of allowing autonomy in vehicles in the automotive industry, which has shown a lot of interest in this technology. There are many advantages to autonomous vehicles, including reduced emissions due to traffic flow, increased utilization due to car fleets, and increased protection due to less human errors. This technology provides a highly lucrative area of interest and expertise for Sigma Technology and its clients, as one of Sweden's leading consulting firms of consultants who work closely with the automotive industry. As a result, an exposure to autonomous technology research and the safety standards for vehicles that use the technology is a very useful asset.

The main goal is to create a fully automated system that uses autonomous logistic transportation by using:

• Techniques for delivering self-driving capabilities using image processing;
• Tracking the location of goods using GPS.

13.2 LITERATURE SURVEY

The information gathered from the literature survey below can be used in a variety of ways to progress the work. Similarly, each paper has its own set of advantages and disadvantages, as well as suggestions for overcoming the disadvantages. A review of state-of-the-art datasets and techniques on Computer Vision for Autonomous Vehicles is presented by Joel, Fatma, and Aseem [1]. It covers both historical and current literature on a variety of subjects, including identification, restoration, motion estimation, monitoring, scene comprehension, and end-to-end learning. Dong, Xiao-Yang, and Xiao [2] have discussed how to improve the safety of self-driving vehicles using computer vision methods and to develop a control strategy for decision making and to reduce the cost of effective multi-lane radar. Giuseppe [3] discusses the personification of car intelligence, which includes an algorithmic "brain," a simulated human "voice," and strong sensor-based "senses." In his Doctor of Philosophy dissertation, Zhilu [4] explored computer vision and machine learning algorithms, as well as their implementations for autonomous vehicles. He has also emphasized the use of end-to-end learning to propose an autonomous lane-keeping method. Sun [5] proposed a faster operation for transporting goods, incorporating GPS and RFID.

In order to make transportation autonomous we employ Computer-Vision algorithms. In this methodology, we use Raspberry pi (master device), Arduino (Slave device), and Raspberry pi cam (Capturing images). Here Open CV library is used for image processing operation, which will be discussed in subsequent sections. The main purpose of this work is to effectively use Computer-Vision technique in autonomous vehicles (Cargo transportation) and to minimize the labor cost and unsafe behavior in the transportation vehicle.

13.3 HARDWARE COMPONENTS

In this section, we are going to discuss about the components that are being used in the Autonomous Transportation System project and how each component works.

13.3.1 RASPBERRY PI 3 MODEL B+

This application is served by a Raspberry Pi 3 Model B+ with a 64-bit quad-core processor operating at 1.4 GHz and dual-band 2.4 GHz and 5 GHz wireless LAN. The wireless LAN with dual-band has modular compliance certification that remarkably reduces compliance testing of wireless LAN. This improves the time to market and cost. This acts as the brain of the system which predicts the direction of the vehicle using the machine learning algorithm proposed in this system (Figures 13.1 and 13.2).

FIGURE 13.1 Raspberry PI 3 Model B+.

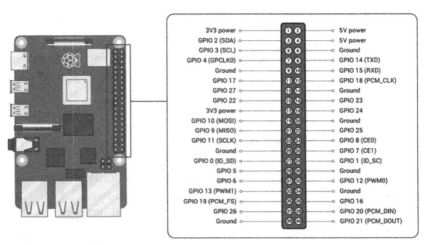

FIGURE 13.2 Pin diagram of raspberry PI 3 model B+.

13.3.2 RASPBERRY PI CAMERA MODULE

The Raspberry Pi Cam consists of a 5 MP CMOS camera and a fixed focus lens that can capture still images as well as high-definition videos. The resolution of the stills are 2592 x 1944; 1080 p at 30 FPS, 220 p at 60 FPS, and 640×480 at 60 or 90 FPS video works. Raspberry Pi used operating system. It is used to capture images (Figure 13.3).

FIGURE 13.3 Raspberry Pi camera module.

13.3.3 ARDUINO

The Arduino platform seems to be a free and open-source electronics platform. It has simple hardware and software. A series of instructions can be written on the microcontroller using the Arduino programming language, which is based on Wiring, and the Arduino Software (IDE). Based on instruction received from raspberry pi it will drive the motor. It acts as a slave device. A Parallel mode of communication is employed between Raspberry Pi and Arduino.

13.3.4   SERVO MOTOR

A servomotor is a special electrical motor to rotate objects with higher precision. Velocity, acceleration, and angular or linear position are controlled by servomotor. The position feedback is provided by encoder or resolver. The term servomotor refers to a motor suitable for usage in a closed-loop control system. Servo Motors are used to lift the cargo and place it in the storage system in the vehicle.

13.3.5   ULTRASONIC SENSOR

Ultrasonic sensors consist of a trigger pin and an echo pin. It works by emitting sound waves of high frequency and so, it waits for the sound to urge reflected. Supported the time taken for reflection, it calculates the distance of the target object. In this methodology, the trigger pin emits sound waves of high frequency and the echo pin receive the reflected sound. We employ ultrasonic sensors to detect objects that are of greater height. Speed breaker comes under low-level height, and it wouldn't be detected as an object. In order to enhance accuracy HAAR cascade classifier was used to detect lower height objects (Figure 13.4).

FIGURE 13.4   Ultrasonic sensor.

13.4 LIBRARY USED

13.4.1 OPEN CV

Written in C and C++, this open-source computer vision library runs on Linux, Windows, and Mac OS X. It has Python, Ruby, MATLAB, and other language interfaces. The main goal of OpenCV is to have high computational efficiency and solve complex real-time problems. OpenCV offers a simple architecture that allows people to easily create more complex applications. OpenCV has more than 2,500 optimized algorithms. Applications of OpenCV include facial recognition, traffic sign detection; tracking the movement of objects, vehicle detection, etc.

13.4.2 PERSPECTIVE TRANSFORM

In this approach, perspective transform has been used in order to gain a better understanding of the necessary detail. We need to include the points where we need to collect information from a picture by shifting perspective in order to get information from it. It is also important to include the internal points where we want our picture to be shown. We get the perspective transform from the two sets of points. After that it is wrapped up with the original image.

1. **getPerspectiveTransform():** Birds-eye view of a picture is obtained from this. It is also accustomed to calculate perspective transform from four pairs of the points obtained from Region of Interest (ROI). Region of Interest is obtained from the original image is in the shape of a polygon, where actual image processing takes place.
   Representation matrix of a perspective transform so that:

$$\begin{bmatrix} t_i x_i' \\ t_i y_i' \\ t_i \end{bmatrix} = map_{matrix} \cdot \begin{bmatrix} x_i \\ y_i \\ 1 \end{bmatrix} \quad (1)$$

where;

$$dst(i) = \left(x_i', y_i'\right), src(i) = \left(x_i, y_i\right), i = 0,1,2,3 \qquad (2)$$

2. **WarpPerspective:** In order to align the coordinates of the frame Warp perspective is employed, only after getting results of the perspective transform. Mathematical representation of warp Perspective that transforms source image using the specified matrix:

$$dst(x,y) = \text{src}\left(\frac{M_{11}x + M_{12}y + M_{13}}{M_{31}x + M_{32}y + M_{33}}, \frac{M_{21}x_1 + M_{22}y + M_{23}}{M_{31}x + M_{32}y + M_{33}}\right) \qquad (3)$$

13.4.3   IN RANGE

We used the in range function to perform threshold operation to distinguish lanes on roads. The in range feature is used after the image has been converted to grayscale from RGB.

13.4.4   CANNY EDGE DETECTION

An edge detection technique's main goal is to determine the boundaries of objects in an image. In this methodology, after removing noise in an image, it's necessary to compute gradient intensity representation of an image. Then we need to use a threshold filter to get lower and upper boundary on gradient values. Each pixel represents the light intensity at a specific point in the image. The pixel's intensity is represented by a numeric value ranging from 0 to 255, with zero indicating a black surface. The value 255 denotes the highest density of something that is fully white. In the following step, we need to trace down the hysteresis by suppressing weak edges that aren't connected to strong edges (Figures 13.5 and 13.6).

13.4.5   HISTOGRAM OPERATION

After obtaining the results of the threshold operation made on the image (in range + canny edge), we must locate the pixels that belong to the left lane and those that belong to the right lane. In this methodology, we're making a histogram of the image's lane line pixels. In order to locate lane position, we employ histogram operation. Two examples of histograms produced from a binary image are in Figure 13.7.

FIGURE 13.5 Original image.

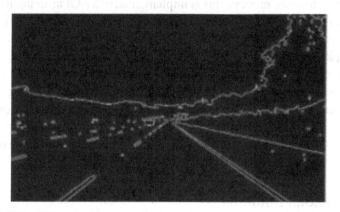

FIGURE 13.6 Edge detection by canny.

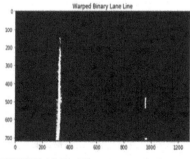
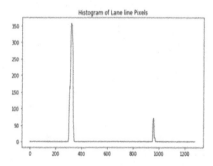

FIGURE 13.7 Histogram operation.

The two peaks in the picture above represent the left and right lanes, respectively. Then we need to scan images within these two-lane positions, adding detected pixels to a vector beginning at the bottom and working up to the top of the image. The lane's Mid-point is derived based on the position of the left and right lanes.

The car's position in relation to the lane is determined. The location of the two observed lane lines, as well as the position of the camera, which is assumed to be in the image's middle, are used to determine the car's relative position to the lane.

13.4.6   STOP SIGN AND OBJECT DETECTION

The Haar-Cascade classifier is trained with hundreds of positive samples of a selected object (such as a stop sign, object, or traffic light), as well as negative samples. These random images are the same size. After completion of the training process, this is implemented to a ROI in an input picture. If area is likely to display the object (i.e., stop sign, object traffic light), the classifier outputs "1," otherwise "0." In the entire picture, the classifier looks for a specific object or Stop sign. The classifier has a property of resizing, so that it finds the objects at different sizes. Hence it avoids resizing images, which is considered to be more efficient. Iterating the scan procedure several times at different sizes will reveal the unknown-size object in the image. We then calculate the distance between the vehicle from the stop sign, object, and traffic light using the straight-line equation:

$$y = mx + c \tag{4}$$

13.5   BLOCK DIAGRAM

For lane detection (Figure 13.8):

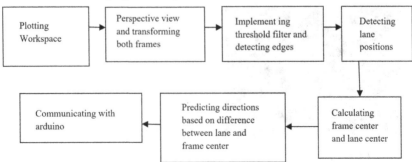

FIGURE 13.8   Block diagram.

13.6 WORKING OF HARDWARE

In this method, the Pi cam is used to capture photographs. I2C is used to feed images from the Pi cam to the Raspberry Pi. Computer vision methods are then used to process the images. The Raspberry Pi sends commands to Arduino based on the computer-vision algorithm's predictions. Based on signal, the motor driver receives pulse from Arduino. Once you've arrived at your destination, Arduino will take action (Figure 13.9).

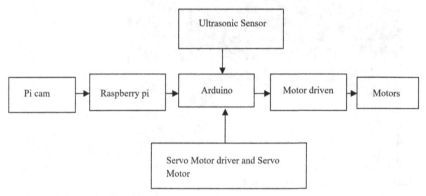

FIGURE 13.9   Working of hardware.

13.7 HARDWARE MODULE (FIGURE 13.10)

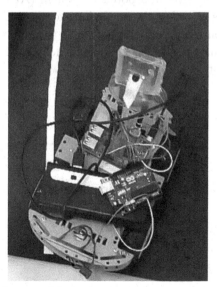

FIGURE 13.10   Hardware module.

13.8 RESULT

On the output screen, the resultant image generated after applying the lane detection algorithm and Haar-Cascade classifier on the stop sign is displayed (Figure 13.11).

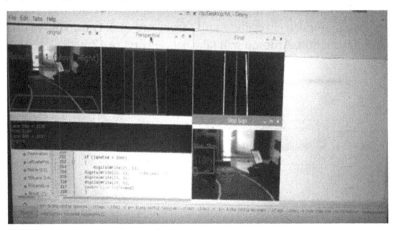

FIGURE 13.11 Resultant image obtained.

13.9 CONCLUSION

This autonomous logistic transportation system (ALTS) can be used in healthcare for transporting the medicines to the patients. During this COVID-19 pandemic where social distancing and safety of health workers is most important, we can use this ALTS to provide medicine to the COVID-19 patients.

KEYWORDS

- Arduino
- autonomous logistic transportation system
- COVID-19
- deep learning
- Haar-Cascade classifier
- lane detection algorithm

REFERENCES

1. Joel, J., Fatma, G., & Aseem, B., (2020). *Computer Vision for Autonomous Vehicles: Problems, Datasets, and State of the Art.* Publisher: Now Foundations and Trends.
2. Dong, D., Xiaoyang, L., & Xiaotong, S., (2018). A vision-based method for improving the safety of self-driving. In: *2018 12th International Conference on Reliability, Maintainability, and Safety (ICRMS).*
3. Giuseppe, L., (2017). Virtual assistants and self-driving cars. In: *2017 15th International Conference on ITS Telecommunications (ITST).* Warsaw, Poland.
4. Zhilu, C., (2017). *Computer Vision and Machine Learning for Autonomous Vehicles.* A Dissertation Submitted to the Faculty of the Worcester Polytechnic Institute in partial fulfillment of the requirements for the Degree of Doctor of Philosophy in Electrical and Computer Engineering.
5. Sun, J., (2012). Design and Implementation of IoT-Based Logistics Management System. Published in 2012. *IEEE Symposium on Electrical and Electronics Engineering (EEESYM).*
6. Sun Jianli (2012). "Design and Implementation of IOT -Based Logistics Management System." Published in 2012. *IEEE Symposium on Electrical & Electronics Engineering (EEESYM),* (24–27 June 2012).
7. Dong Dong, Xiaoyang Li, & Xiaotong Sun, (2018). "A Vision-based Method for Improving the Safety of Self-driving." Published in 2018. 12th International Conference on Reliablity, Maintainability, and Safety (ICRMS), (17-19 Oct. 2018).

CHAPTER 14

Survey on Cancer Diagnosis from Different Tests and Detection Methods with Machine and Deep Learning

GHOST MANOJ KUMAR,[1] P. SATHISH KUMAR,[2] and V. RAJENDRAN[3]

[1]Department of Electronics and Communication Engineering, Kallam Haranadha Reddy Institute of Technology, Chowdavaram, Guntur, Andhra Pradesh, India, E-mail: Manojkumar.ece2016@gmail.com

[2]Department of Electronics and Communication Engineering, Bharath Institute of Higher Education and Research, Selaiyur, Chennai-73, Tamil Nadu, India, E-mail: sathishmrl30@gmail.com

[3]Department of Electronics and Communication Engineering, Vels Institute of Science, Technology, and Advanced Studies (VISTAS), Chennai, Tamil Nadu, India

ABSTRACT

Cancer is one of the most dreaded and antagonistic diseases in the world, being meticulous in excess of 9 million deaths unanimously. Many researchers established that detecting cancer disease at early stages leads to an increase in probability to recuperate from cancer disease. In this proposed literature review work, we have chosen machine learning and also deep learning methods for cancer disease diagnosis in human body through algorithm based on issues, laboratory tests, imaging tests, biopsy, bone scan, computerized tomography (CT) scan, positron emission

Application of Deep Learning Methods in Healthcare and Medical Science.
Rohit Tanwar, PhD, Prashant Kumar, PhD, Malay Kumar, PhD, & Neha Nandal, PhD (Editors)
© 2023 Apple Academic Press, Inc. Co-published with CRC Press (Taylor & Francis)

tomography (PET), ultrasound images, etc. Moreover, we investigated how existing work analyzed cancer disease using recent technologies. The main objective of this state-of-art work is to evaluate performance by accuracy estimation on cancer disease prediction via machine learning. Furthermore, comparison of existing work specifies that how cancer disease were diagnosed, ways to prevent during early stage of cancer disease, or classification of disease by means of deep learning approaches were scrutinized, which makes cooperative in medical diagnosis synchronal relevance. The intention of this state of artwork is to afford investigators to choose to work in realizing machine learning as well as deep learning approaches for cancer disease identification.

14.1 INTRODUCTION

Cancer is the abandoned expansion of anomalous cells in the human body. Suddenly, the standard (normal) control mechanism stops functioning once cancer grows. Formation of new cells or broaden of cells in human body and also existing cells do not destroy, that extension of cells are referred as tumor. A few types of cancer doesn't appear tumors as leukemia. Cancer may possibly arise anyplace in the human body. In worldwide, probably ordinary men and women were affected by prostate cancer and breast cancer disease.

Commonly, cancer has divided into five categories:

1. **Carcinomas:** Cancer arises either in skin or cells or tissues of internal limb.
2. **Lymphomas:** Cancer occurs in body immune system.
3. **Leukemia:** Enlarging of blood cells and bone marrow.
4. **Sarcomas:** Cells grow beyond in connective tissues.
5. **CNS (Central Nervous Systems):** Arises in spinal cord, brain.

The cancer disease is being treated as follows:

- **Step 1:** Medicinal individuals chose *surgery* to remove the tumor completely.
- **Step 2:** The cancer cells are destroyed by using chemicals called *chemotherapy*.
- **Step 3:** Also, cancer cells can be destroyed by using X-ray called *radiotherapy*.

Nowadays, modern technologies such as machine learning and deep learning (DL) algorithms are utilized in every application domain, especially in medicinal fields such as identification and investigation of disease to predict about future assessments. Hence, we proposed the state of artwork that indicates early cancer disease detection and techniques were developed to prevent cancer by various investigators, and also how the disease were distinguished from normal while performed on images as CT, US, PET, etc., had done by various researchers work using machine learning and DL techniques.

The main objective of this review work is as follows:

- to study various machine learning algorithms in detecting cancer disease such as breast cancer, lung cancer and skin cancer;
- to analyze how DL algorithms are applicable in finding abnormalities in human blood cells which may be either breast or skin or some other parts of the body;
- to compare machine learning and DL based NN algorithms to recognize which approaches are suitable for cancer disease diagnosis;
- to discuss about metrics evaluation such as sensitivity, specificity, f-score, and precision;
- to focus on comparing the accuracy measures on various algorithms (ML and DL) to evaluate the overall performance of algorithm in disease prediction;
- to examine several architectures of CNN algorithm developed by several researchers to highlight the cancer disease prediction, which is helpful in medicinal applications.

14.2 SEVERAL CANCER DISEASE PROGNOSIS USING DEEP LEARNING TECHNIQUES

1. **Breast Cancer:** Nan et al. [1] datasets comprise of 229,426 digital screening mammography images were gathered from many patients in hospitals. The researcher built deep CNN for diagnosis breast cancer via screening, categorizing disease, training, assessment, and generate outcome as separation of normal and abnormal (cancer cells enlarges on breast) cells. Jeffrey et al. [2] developed Artificial techniques specifically DL algorithms to detect lymph nodes available in breast enlarged cells on digital images.

Padideh et al. [3] developed DL algorithm for prognosis cancer disease also decisive genes recognition for breast cancer prognosis. The relevant features extracted from very high dimension DNA appearance patient profile using Stack Denoising Auto Encoder (SDAE). This interactive genetic material might be helpful cancer biomarkers for breast cancer diagnosis applicable in medicinal domains.

Babak et al. [4] applied DL techniques in finding metastases on hematoxylin as well as regions in lymph nodes of women who is having breast cancer. A comparison was also made between novel outcomes with the analysis done by pathologists.

Naresh et al. [5] gathered WBCD (Wisconsin Breast Cancer Database) for verdict breast cancer patients using DL knowledge which achieved accuracy of 99.67%. This work has three stages as:

i.   **Pre-Processing Phase:** Label encoder, standard scalar, normalize scalar.
ii. **Training Data:** Data has been training.
iii. **Testing Data:** Perform validation to generate outcome in predicting breast cancer.

Also, comparisons have been made in DL algorithms along with machine learning methods in predicting breast cancer disease. Finally, DL methods outperforms in achieving greater performance through accuracy estimation in cancer disease diagnosis. During pre-processing phase, normalized scalar is being used to rescaling the features that range from 0 to 1. The formula 1 is to normalize within specified range as Eqn. (1):

$$\frac{x_i}{\sqrt{x_i^2 + y_i^2 + z_i^2}} \tag{1}$$

Also, standard scalar methods were utilized to rescaling the features during pre-processing stage. Eqn. (2) expressed to predict the standardization while dataset has Gaussian distribution.

$$\frac{x_i\text{-mean}(x)}{\text{stdev}(x)} \tag{2}$$

Yusuf et al. [6] the most widespread division of breast cancer (i.e.), Invasive Ductal Carcinoma was identified automatically using DL algorithms as ResNet 50, and DenseNet 161. Metrics such as sensitivity, specificity, precision, f-score, and BAC were found to evaluate overall model performance in predicting breast cancer. Performance calculated chart using ResNet 50 and DenseNet 161 as shown in Table 14.1.

TABLE 14.1   Chart Shows Performance Estimation on DL Techniques

Metrics	Performance Calculation	
	ResNet 50	DenseNet 161
Sensitivity	93.6%	89.6%
Specificity	88.3%	93.6%
F-score	94.1%	92.4%
Precision	94.6%	95.3%
Accuracy	91.95%	91.2%

Li et al. [7] predicted breast cancer disease on screening mammography (FFDM) images using DL algorithm specifically CNN. The greatest solitary method reaches accuracy as 88%, and then they analyzed for models thereby accuracy increases to 91% is detailed in Table 14.2.

TABLE 14.2   Accuracy Comparison between Single Model and Four Models

	Accuracy for Solitary Model	Accuracy for Four Models
CBIS-DDSM	88%	91%
FFDM	95%	98%

Sebastien et al. [8] exposed earlier period of 20 years that various techniques were used especially mammography images were utilized for prognosis of breast cancer most commonly affected by women. The limitations of using mammography images as not correctly diagnosed the cancer disease means false positive (FP) of mammography images may happen that the patient is established positive by some other methods. Hence, this work investigated infrared digital imaging, thermal contrast [9], among breast cancer patient with normal breast reveals enhancement in thermal action in premature cancer cells and regions circumstances enlarges breast cancer. This survey recognized CAD tools along with IR image for finding cancer disease using DL approaches. If self-assurance production of deep neural network (NN) is higher than 0.5 and least value as 0.6 using support vector to categorize normal breast from breast cancer with thermal image (Table 14.3). Total data's utilized as healthy and disease affected patients under training and testing.

TABLE 14.3 Data Usage Description by Amria et al. [9]

	Training	Testing	Overall
Normal	481	121	602
Patients Affected by Cancer	368	92	460
Overall	849	213	1,062

For model evaluation, this work utilized 12 breasts to predict and classify normal from cancer disease affected patients.

2. **Lung Cancer:** Wenqing et al. [10] analyzed lung cancer disease on LIDC (Lung Image Database Consortium) using DL techniques such as CNN, DBN, SDAE. The input CT images were divided in accordance with the marker region afforded by a physician. Comparing performance of various DL algorithms shown in Table 14.4.

TABLE 14.4 Performance Estimation Based on Accuracy by Wenqing, Bin, and Wei [10]

Deep Learning Technique	Accuracy
CNN	80%
DBN	81%
SDAE	79%
CADx	79%

3. **Skin Cancer:** Mohammad et al. [11] scope is to categorize epidermal cell images from normal cells to find skin disease by developing DL techniques. Also, accuracy estimation had found to enhance the performance in classifying skin cancer from others via constructing model-driven DL frameworks.

4. **All Cancer Diseases Prognosis:** Alaá et al. [12] examined many Artificial Intelligent (AI) techniques to diagnose various cancer disease as lung, breast, and brain cancer using various images such as US images, Magnetic Resonance Images, X-ray images

and Computerized Tomography (CT). Finally, performance was evaluated to discover which algorithm outperforms in predicting cancer disease.

Rasool et al. [13] improved this work from existing work regarding cancer diagnosis methods makes opportunity of gathered information from several types of cancer via feature mining techniques that facilitate to improve recognition and identification of disease. But this work introduced novel technique for cancer prognosis and categorizing cancer varieties founded on gene expression data. Also, the performances were evaluated by comparing various applied algorithm for cancer detection and analysis. To estimate the strength of classifier, 10 fold cross-validation have been presented based on classification accuracy. Comparison had also done for cancer disease detection using SVM along with Gaussian core as well as SoftMax regression.

14.2.1 PREDICTION OF CANCER DISEASE USING MACHINE LEARNING TECHNIQUES

1. **Breast Cancer:** Abien et al. [14] gathered data's from source as WDBC to predict breast cancer disease using machine learning algorithms such as support vector machine (SVM), linear regression (LR), multilayer perceptron (MLP), Neural Network (NN), SoftMax regression. After gathered breast cancer related datasets, undergoes pre-processing to extract relevant features only using the following formula 3.

$$Z = \frac{X - \mu}{\sigma} \tag{3}$$

Machine learning algorithms were built to predict cancer disease and perform classification algorithms to categorize cancer disease from normal. The data were analyzed via training and testing phase, also metrics evaluation, namely True positive, True Negative, False Negative, FP, accuracy, and epochs to validate overall performance in detecting breast cancer disease.

Mohd. Rasoul et al. [15] put forth early diagnosis of cancer disease that makes patients life secure. This researcher investigates many mammography images to identify carcinoma arises in the breast using image processing and machine learning algorithm such as logistic regression and Backpropagation NN. Habib et al. [16] distinguished normal from abnormal (breast cancer) via genetic programming as well as machine learning algorithms. Feature extraction had completed using genetic programming. Also, performance in detecting breast cancer and classification were carried out via metrics evaluation, namely sensitivity, recall, precision, accuracy, and Region of Curves which generates the best model. Meriem et al. [17] proposed machine learning algorithms such as KNN and NB for classifying breast cancer from normal. The breast cancer is specified as abnormal, which is identified using features such as size modification, red color, continuous pain, genes transformation, etc. Finally based on accuracy estimation, the algorithm performance in detecting breast cancer was identified.

Sri Hari et al. [18] implemented that machine learning algorithms such as logistic regression, KNN, SVM for classifying normal (benign) as non-cancer and malicious (breast cancer).

Sreyam et al. [19] developed machine learning algorithms such as logistic regression, Naïve Bayes (NB), NNs, and decision tree (DT). In addition to that, the prognosis on breast cancer was analyzed depending on parameter extraction. With the more level of accuracy, the performance of the algorithm was estimated in detection of breast cancer. Figure 14.1 depicts the overall workflow of this anticipated model by Sreyam, here, which machine learning algorithms applicable for prognosis cancer detection via finding accuracy along with its performance.

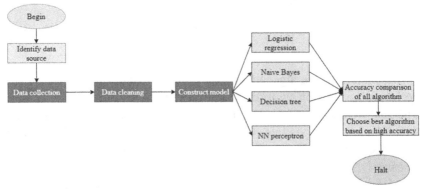

FIGURE 14.1   Workflow for anticipated model by Sreyam, Ronit, and Swarnalatha [19].
Source: Redrawn from Ref. [19].

Tahmooresi et al. [20] found that mammogram images like X-ray photograph, X-ray print photograph, etc., may threat of disease detection not accurate occasionally leads to imperil patients health. To solve such kind of issues as diagnosis disease prognosis very accurately this work presented hybrid model which comprises of combination of Support vector machine, ANN, k-nearest neighbor (KNN), and DT for successful breast cancer prognosis using dataset with images and blood test clinical reports too.

Shahnorbanun et al. [21] projected two machine learning tools as CADe depends on SVM algorithm and CADx ensemble of four Radial Based classifiers. Both proposed works suggested unconventional decision making capability that helps in the medical domain on predicting cancer disease. Therefore, it decreases FP rate, which basis over-segmentation also enhances overall performance in finding and prognosis of breast cancer.

Shubham et al. [22] aspired to initiate relationship among various machine learning methods such as KNN, NB, RF frequently utilized for breast cancer prognosis.

The dataset has been taken from WDBC (Wisconsin Diagnosis Breast Cancer) to detect cancer disease and handling disease by estimating key metrics as precision, accuracy, sensitivity, specificity, recall. The accuracy predicted for KNN, NB, and RF as 96%, 94.5%, and 94.7%. Metrics were estimated using formula exposed in Table 14.5.

TABLE 14.5 Metrics Along with Specific Formula by Sharma, Aggarwal, and Choudhury [22]

Metrics	Equation
Accuracy	
Precision	
Recall	
F1-score	

Hiba et al. [23] determined that machine learning algorithms such as SVM (accuracy as 97.13%) outperformed in the detection of breast cancer disease and analysis had been made to prevent the risk of breast cancer disease accomplished experimentation on (Wisconsin) breast cancer datasets. This chapter surveyed four machine learning algorithms and comparison have been done to identify cancer disease especially breast cancer via accuracy estimation as well as metrics evaluation.

2. **Lung Cancer:** Kanchan et al. [24] introduced novel techniques as machine learning algorithms along with the Internet of Things in which devices related to medical fields for disease identification, especially lung cancer. Radhika et al. [25] reviewed several studies for predicting lung cancer using machine learning classification methods explicitly decision tree (DT), logistic regression, Support vector machine and NB algorithm. The overall contribution of the chapter is the premature prognosis of lung cancer disease by calculating the performance of classification accuracy. The accuracy estimation by analyzing lung cancer dataset is specified in Table 14.6.

TABLE 14.6   Accuracy Calculation Using ML Techniques [25]

Techniques	Accuracy in Percentage (%)
Decision tree	90%
SVM	99.2%
Logistic regression	67%
Naïve Bayes	87.9%

3. **Prostate Cancer:** Lal Hussain et al. [26] premature recognition of cancer disease may successively decrease the death rate caused by prostate cancer. Owing to the very high resolution of Magnetic Resonance Images of prostate cancer needs appropriate problem-solving systems and tools. In earlier period, many researchers built CAD (Computer-Aided Diagnosis) tools that facilitate radiologist to distinguish anomalies. This work mainly focused on prostate cancer prognosis using machine learning algorithms like SVM, Radial Base Function, DT, and Gaussian. Here, several feature mining approaches as texture-based, morphology operations, SIFT, and EFD features were introduced to enhance prostate cancer detection performance.

The overall techniques used for detecting prostate cancer disease via accuracy estimation as described in Table 14.7.

TABLE 14.7   Accuracy Estimation Using ML by Lal et al. [26]

Method	Accuracy
SVM Gaussian	98.3%
SVM Gaussian with texture + morphology	99.7%
EFD+ morphology features	100%

Figure 14.2 depicts the overall framework in detecting prostate cancer disease using attribute mining approaches such as EFD, entropy, SIFT, texture, and morphological operations. Finally, 10 cross-fold validation was performed to measure accuracy rate in prognosis of prostate cancer very faultlessly.

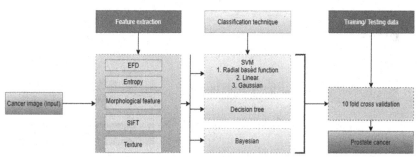

FIGURE 14.2   Overall structure in prognosis of prostate cancer.

4. **Oral Cancer:** Sandhya et al. [27] applied several machine learning algorithms such as DT, regression tree, instance-based, clustering, Bayesian, ANN, DL, dimensionality reduction algorithm for detection of oral cancer and diagnosis had done.

Madhura et al. [28] introduced a novel algorithm namely SVM to identify oral cancer and distinguished normal from abnormal via Apriori algorithm. Moreover, data mining (DM) and extraction were built to retrieve relevant information and association rule from Apriori also constructed to distinguish association among features of oral cancer disease.

The stages of cancer disease had found using SVM technique by finding accuracy along with its metrics as Table 14.8.

TABLE 14.8 Metrics Calculation Using SVM by Madhura et al. [28]

Method	Precision	Sensitivity	Specificity	Accuracy
SVM	95%	95%	96%	97%
Mathura [28]				

Prabhakaran et al. [29] Over intake of smoking and tobacco are crucial threat issue for oral cancer development. The 100 input images have taken and processing undergoes to remove irrelevant features, apply segmentation to distinguish parameter possessions available in images and marker the affected region makes easily discriminate normal and oral cancer in cavity.
The infrastructure for predicting oral cancer is depicted in Figure 14.3.

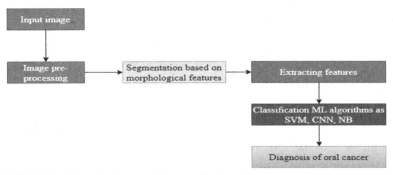

FIGURE 14.3 Overall framework for prediction of oral cancer by Prabhakaran and Mohana [29].

5. **Skin Cancer:** Vijayalakshmi et al. [30] determined skin cancer disease automatically on lesion images collected from medicinal clinical reports. It comprises of three stages a. Data collection and expansion b. Build model c. Prediction. This researcher analyzed various Artificial Intelligent model especially CNN and machine learning as combination of SVM and image processing to build exact structure, which leads to greater accuracy as 85%. Sanjana et al. [31] diagnosed skin cancer disease using machine learning algorithms like LR, logistic, polynomial, stepwise, ridge, lasso, elastic net regression without any human involvement. Because earlier works had been developed manually.

Also, by Sanjana [31] work defines about skin cancer stages:

- Stage 0: Tumors arises in skin but not entered into. This means tumors occur only in skin epidermis.
- Stage 1: Tumor goes underneath epidermis but enters into the next layer of skin called dermis.
- Stage 2: Tumors looks very big nearly 1 mm thickness leads to high risks.
- Stage 3: Melanomas expanded.
- Stage 4: Many different sizes of tumors.

14.2.2   STATE-OF-ART ON CANCER DIAGNOSIS

This review chapter focused on artificial intelligence (AI), especially deep NNs might be profitably utilized for analyzing images [32]. The vital structure of this review is how machine learning algorithms are helpful in detecting medicinal domains such as detecting disease via several phases as preprocessing, image segmentation, and finally post-processing. Moreover, this study analyzed how DL techniques particularly CNN, and various autoencoders, Recurrent Neural Networks (RNN), and Long Short Term Memory. Finally, this study scrutinized DL algorithms for cancer disease (skin, breast, brain, and lung) diagnosis by python code implementation.

The steps for diagnosing cancer disease formulated by Khushboo et al. [32] as follows:

1. **Pre-Processing:** Reduction of irrelevant information about images to enhance quality.
2. **Image Segmentation:** The chosen input image is categorized into several segments for auxiliary processing to continue.
3. **Post Processing:** Final processing to categorize abnormal region from normal areas.

Wan et al. [33] demonstrated how DL approaches useful in medical applications, especially cancer disease prediction. Also, the investigator proved that DL techniques helpful in diagnosis of cancer disease while functioning with large amount of data.

Joseph et al. [34] defines the capability of machine learning models which are applicable in medical prognosis especially cancer disease. This study exposed sketch out of DT which can be utilized for breast cancer disease prognosis and handling disease is described as Figure 14.4.

Basically, DT construction specified by many learners utilized to categorize the decision like 'yes' or 'No,' 'normal' or 'abnormal' likewise. A DT may be erudite by gradually dividing the labeled training data into subsets depends on an arithmetical or rational test [35].

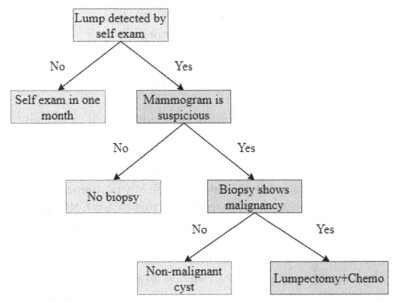

FIGURE 14.4   Breast cancer detection using decision tree.

Ripley, Harris, and Tarassenko [36] introduced NN technique for cancer prediction and Inês Domingues et al. [37] offered a reviewed paper of DL techniques being used in two various image perceptual such as CT, PET, and also amalgamation of both perceptual which has been a significant milestone in numerous assessments depends on plentiful diseases. The author considered around 180 related works available from the year 2014 to 2019, especially for image perceptual.

Also, Ines (2019) survey portrays the involvement of DL methods to several basic roles and responsibilities in medicinal image psychoanalysis, segmentation, classifying normal and abnormal from images, and finally spotlighted two various images as CT and PET.

Kristy et al. [38] analyzed several researchers study primarily focused on the production of precious secretome records. The pie chart allotment explains the types of cancer disease related with secretome records depicted in Figure 14.5.

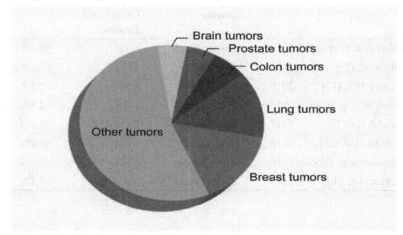

FIGURE 14.5   Pie chart representation on types of cancer.

Gayathri et al. [39] reviewed several papers as state of the art in prognosis of cancer disease using machine learning technique along with enhancement in accuracy estimation of breast cancer disease which leads to decease rate diminution. This work surveyed NN algorithms, Relevance vector machine and also SVM in the prediction of breast cancer disease.

Zilong et al. [40] surveyed numerous research papers in all types of cancer disease prognosis such as brain cancer, breast, skin, lung, prostate, cervical, liver, bladder, colonial cancer. Moreover, concern DL models as CNN, fully CNN, Autoencoders, DL based belief networks were utilized in cancer disease prognosis from images such as US, CT, X-ray, and PET.

14.2.3   COMPARISON EVALUATION AMONG SEVERAL SURVEYS

This survey exposed the overall comparison on different surveys for performance findings using machine learning algorithms and DL algorithms by accuracy estimation in percentage along with related metrics like sensitivity, specificity, recall, and precision specified in Tables 14.9 and 14.10.

TABLE 14.9   Comparisons on Various Surveys Using ML

Survey	Year	Kind of Cancer Disease	Algorithm Gives Good Results	Accuracy
Abien et al. [14]	2019	Breast cancer	MLP	99.04%
Hiba et al. [23]	2016	Breast cancer	SVM	97.13%
Meriam et al. [17]	2018	Breast cancer	KNN	96.19%
Mohd. et al. [15]	2016	Breast cancer	KNN	93.8%
Padideh et al. [3]	2017	Breast cancer	SVM-RBF	98.3%
Radhika et al. [25]	2018	Lung cancer	SVM	99.2%
Shubham et al. [22]	2018	Breast cancer	KNN	95.9%
Sreyam et al. [19]	2019	Breast cancer	NB	98.4%

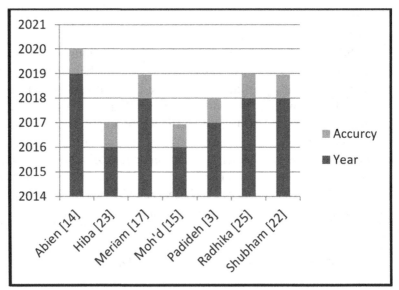

FIGURE 14.6   Accuracy comparison among various surveys in bar chart representation.

In today's world, women are commonly affected by breast cancer. To detect several types of cancer disease such as skin cancer, lung cancer, oral cancer including breast cancer, many researchers developed machine learning and DL techniques. In this proposed review, the comparison made among several researchers detected cancer disease via different techniques depicted in Figures 14.6 and 14.7 as bar chart representation.

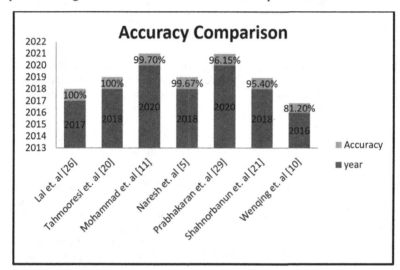

FIGURE 14.7 Accuracy comparison among several surveys using DL.

TABLE 14.10 Comparisons on Various Surveys Using DL

Survey	Year	Kind of Cancer Disease	Algorithm Gives Good Results	Accuracy
Lal et al. [26]	2017	Prostate cancer	EFD+ morphological features	100%
Tahmooresi et al. [20]	2018	Breast cancer	ANN	100%
Mohammad et al. [11]	2020	Skin cancer	Squeeznet	99.7%
Naresh et al. [5]	2018	Breast cancer	CNN	99.67%
Prabhakaran et al. [29]	2020	Oral cancer	CNN	96.15%
Shahnorbanun et al. [21]	2018	Breast cancer	Rejection model $(13 \times 13 \times 13 \times 13)$	95.4%
Wenqing et al. [10]	2016	Lung cancer	DBN	81.2%
Yusuf et al. [6]	2020	Breast cancer	DenseNet161	91.6%

TABLE 14.11   Overall State-of-Art Counter on Various Research Works for Proposed Work

Research Title	Data Source	Specification	Techniques Used	Scope
Nan et al. [1]	1,001,093 breast cancer images	Breast cancer screening	Deep convolutional neural networks	Enhance accuracy in predicting breast cancer which makes helpful for radiologists
Abien et al. [14]	Wisconsin diagnostic breast cancer (WDBC) dataset	Breast cancer detection	Machine learning algorithms as SVM, LR, MLP, NN, GRU-SVM, SoftMax regression	MLP generates better outcomes in predicting breast cancer with accuracy as 99.04%
Kanchan et al. [24]	Previous work	Lung cancer	Machine learning algorithms associated with medical IoT devices	Usage of various machine learning algorithms to detect various diseases along with internet of things
Jeffrey et al. [2]	Digital images	Lymph node in breast cancer	Deep learning algorithms	Detect abnormalities as lymph node in cancer
Habib et al. [16]	NA	Breast cancer	Machine learning approaches	Automatic breast cancer detection
Alaa et al. [12]	US, CT, X-ray, MRI images	Breast cancer, lung, and brain cancer	AI techniques: SVM-NN ANN Fuzzy Adaptive neuro-fuzzy	Examined many algorithms for finding breast, brain, and lung cancer. Also find higher accuracy which produced better outcome in prediction

TABLE 14.11   *(Continued)*

Research Title	Data Source	Specification	Techniques Used	Scope
Mohd. Rasoul et al. [15]	Mammography images	Breast cancer	Logistic regression back propagation neural network	Breast cancer recognition though feature extraction
Meriem et al. [17]	UCI repository breast cancer dataset	Breast cancer classification	1. Naïve Bayes – 96% 2. K-nearest neighbor – 97%	Classifying dataset into normal data and breast cancer as malicious
Sri Hari et al. [18]	Wisconsin breast cancer (WBC)	Breast cancer	Machine learning algorithm as SVM, KNN, LR	Categorizing benign as non-cancer, malignant as cancer
Padideh et al. [3]	–	Breast cancer	DL	Stacked denoising encoder for feature extraction
Tahmooresi et al. [20]	Images	Breast cancer	ML as SVM, KNN, DT, ANN	Early detection of cancer disease to save human's life.
Gayathri et al. [39]	CT scan	Breast cancer	ML	Reviewed many papers for detecting cancer disease
Zilong et al. [40]	Overall survey using deep learning techniques using CNN, fully convolutional network, autoencoders, deep belief networks			
Lal et al. [26]	–	Prostate cancer	Machine learning as Bayesian, SVM, RBF, Gaussian, decision tree	Cancer identification using feature extraction technique
Sreyam et al. [19]	–	Breast cancer	ML-RF, DT, BN, ANN	Cancer prediction using feature selection method
Rasool et al. [13]	Detection of several cancer types using deep learning based on gene expression.			

TABLE 14.11 *(Continued)*

Research Title	Data Source	Specification	Techniques Used	Scope
Sandhya et al. [27]	–	Oral cancer	ML-DT, Bayesian, clustering, instance-based, ANN	Recognition of oral cancer disease occurs in cavity caused by tobacco intake and smoking
Mohammad et al. [11]	Skin cell images	Skin cancer	Deep learning models	Categorize dermal cell images
Babak et al. [4]	129 whole side pathology images	Breast cancer	Deep learning	Detection of lymph nodes along with breast cancer
Naresh et al. [5]	WBDC	Breast cancer	Deep learning	Breast cancer detection by accuracy calculation
Vijayalakshmi et al. [30]	Lesion images	Skin cancer	Image processing and ML	Prediction of skin cancer occurs on epidermis using injury images
Sanjana et al. [31]	Cosmetics, chemicals applied images	Skin cancer	ML and Image processing	Skin cancer detection
Radhika et al. [25]	Images from UCI	Lung cancer	SVM, DT, LR	Early diagnosis of skin cancer
Yusuf et al. [6]	BreakHis breast cancer dataset, Whole slide images	Breast cancer	Deep transfer learning	Detecting persistent ductal carcinoma
Lishen et al. [7]	CBIS, DDSM	Breast cancer	DL	Detect breast cancer via screening mammogram images
Shahnorbanun et al. [21]	Mammography images	Breast cancer	ML-SVM	Segmenting images into normal and cancer using marker watershed controller

TABLE 14.11 *(Continued)*

Research Title	Data Source	Specification	Techniques Used	Scope
Sebastian et al. [8]	Infrared thermal imaging	Breast cancer	DL	Breast cancer diagnosis using infrared thermal images.
Shubham et al. [22]	WDBC	Breast cancer	RF, KNN, NB	Breast cancer detection using ML
Prabhakharan et al. [29]	Features as input	Oral cancer	SVM, NB CNN	Oral cavity detection via parameter extraction technique
Wen et al. [10]	LIDC	Lung cancer	CNN, DBN	Prediction of lung cancer disease on lung cancer images
Hiba et al. [23]	WDBC, SDAE	Breast cancer	SVM, NB, KNN, C4.5 decision tree	Diagnosis of breast cancer disease by accuracy calculation

14.3 CONCLUSION

Nowadays, precise detection and classification of cancer disease is important to make human's life very invulnerable. Hence, our proposed review
analyzed several machine learning algorithms and DL algorithms as
state-of-art to predict cancer disease from mammogram images explicitly
X-ray images gathered from clinical reports or cancer disease datasets.
This evaluation organized to contrast the performance allied with machine
learning and DL techniques in the prediction of several cancer diseases
based on accuracy as well as categorizing the images as normal and
malignant. Moreover, we established that X-ray (mammogram) images
for breast cancer using ANN achieves excellent results with accuracy as
100% in breast cancer prognosis also classification of the same. Furthermore, for prostate cancer EFD along with morphological features reaches
higher accuracy as 100%, for lung cancer disease prediction, support

vector machine yields 99.2% accuracy, CNN generates greater accuracy as 96.15% in oral cancer diagnosis, skin cancer prediction accuracy reaches as 99.7% via Squeeznet algorithm from CNN. Different investigators used different ML and DL techniques for predicting various cancer diseases such as lung, skin, prostate, breast, and oral cancer. This proposed study initiated best cancer detection algorithm through comparative analysis of work done by several investigators on cancer disease diagnosis based on highest accuracy using machine and DL techniques. From this state-of-art, we conclude that both machine learning and DL algorithms is being applicable in cancer disease detection, which is helpful in medicinal domains and medical applications too.

KEYWORDS

- artificial intelligent
- convolutional neural network
- linear regression
- multilayer perceptron
- neural network
- support vector machine
- ultrasound
- Wisconsin diagnostic breast cancer

REFERENCES

1. Wu, N., et al., (2020). Deep neural networks improve radiologists' performance in breast cancer screening. In: *IEEE Transactions on Medical Imaging* (Vol. 39, No. 4, pp. 1184–1194).
2. Golden, J. A., (2017). Deep learning algorithms for detection of lymph node metastases from breast cancer: Helping artificial intelligence be seen. *JAMA, 318*(22), 2184–2186.
3. Danaee, P., Ghaeini, R., & Hendrix, D. A., (2017). A deep learning approach for cancer detection and relevant gene identification. *Pac. Symp. Biocomput., 22*, 219–229.
4. Babak, E. B., Mitko, V., Paul, J. V. D., Bram, V. G., Nico, K., Geert, L., & Jeroen, A. W. M. V. D. L., (2017). Diagnostic assessment of deep learning algorithms for

detection of lymph node metastases in women with breast cancer. *JAMA, 318*(22), 2199–2210.

5. Khuriwal, N., & Mishra, N., (2018). Breast cancer diagnosis using deep learning algorithm. In: *2018 International Conference on Advances in Computing, Communication Control and Networking (ICACCCN)* (pp. 98–103). ISBN: 978-1-5386-4120-0.

6. Yusuf, C., Muhammed, T., Ozal, Y., Murat, K., & Rajendra, A. U., (2020). Automated invasive ductal carcinoma detection based using deep transfer learning with whole-slide images. *Article in Pattern Recognition Letters, 133*, 232–239. ISSN 0167-8655.

7. Li, S., Laurie, R. M., Joseph, H. R., Eugene, F., Russell, M., & Weiva, S., (2019). Deep learning to improve breast cancer detection on screening mammography. *Scientific Reports, 9*, 12495.

8. Sebastien, J. M., Petra, M., Ondrej, K., Ali, S., & Kamil, K., (2018). Breast cancer detection using infrared thermal imaging and a deep learning model. *Sensors, 18*(9), 2799.

9. Amria, A., Pulko, S. H., & Wilk, A. J., (2016). Potentialities of steady-state and transient thermography in breast tumor depth detection: A numerical study. *Comput. Methods Progr. Biomed., 123*, 68–80.

10. Wenqing, S., Bin, Z., & Wei, Q., (2016). Computer-aided lung cancer diagnosis with deep learning algorithms. *Medical Imaging 2016, Proc. of SPIE, 9785*, 97850Z-1.

11. Mohammad, A. K., & Sulaiman Al, R., (2020). Skin cancer detection: Applying a deep learning-based model driven architecture in the cloud for classifying dermal cell images. *Informatics in Medicine Unlocked, 18*, 100282. ISSN 2352-9148.

12. Al-Shamasneh, A. R. M., & Unaizah, H. B. O., (2017). Artificial intelligence techniques for cancer detection and classification: Review study. *European Scientific Journal, 13*(No. 3). ISSN: 1857-7881.

13. Rasool, F., Faisal, L., Azade, N., & Manfred, H., (2013). Using deep learning to enhance cancer diagnosis and classification. In: *Proceedings of the 30ᵗʰ International Conference on Machine Learning* (Vol. 28). Atlanta, Georgia, USA, JMLR: W&CP.

14. Abien, F. M. A., (2019). *On Breast Cancer Detection: An Application of Machine Learning Algorithms on the Wisconsin Diagnostic Dataset.* arXiv, vol abs/:1711.07831v4 [cs.LG].

15. Al-Hadidi, M. R., Abdulsalam, A., & Mohannad, A., (2016). Breast cancer detection using k-nearest neighbor machine learning algorithm. In: *2016 9ᵗʰ International Conference on Developments in E-Systems Engineering* (Vol. 1, pp. 35–39).

16. Habib, D., Eslam Al, M., Awais, M., Wail, E., & Mohammed, F. N., (2019). Automated breast cancer diagnosis based on machine learning algorithms. *Journal of Healthcare Engineering, 2019*, 1–11. Article ID: 4253641.

17. Amrane, M., Oukid, S., Gagaoua, I., & Ensari̇, T., (2018). Breast cancer classification using machine learning. In: *2018 Electric Electronics, Computer Science, Biomedical Engineering's Meeting (EBBT)* (pp. 1–4).

18. Sri Hari, N., Pragnyaban, M., & Suvarna, V. K., (2019). Breast cancer detection using machine learning way. *International Journal of Recent Technology and Engineering (IJRTE)* (Vol. 8, No. 2S3). ISSN: 2277-3878.

19. Sreyam, D., Ronit, C., & Swarnalatha, P., (2019). Feature selection for breast cancer detection using machine learning algorithms. *International Journal of Innovative Technology and Exploring Engineering (IJITEE)* (Vol. 8, No. 9). ISSN: 2278-3075.

20. Tahmooresi, M., Afshar, A., Bashari, R. B., et al., (2018). Early detection of breast cancer using machine learning techniques. *J. Telecommun. Electron. Comput. Eng., 10*(2), 1–27.

21. Sahran, S., Qasem, A., Omar, K., Albashih, D., Adam, A., Norul, H. S. A. S., Abdullah, A., et al., (2018). Machine learning methods for breast cancer diagnostic. In: *Breast Cancer and Surgery.* IntechOpen.

22. Sharma, S., Aggarwal, A., & Choudhury, T., (2018). Breast cancer detection using machine learning algorithms. In: *2018 International Conference on Computational Techniques, Electronics, and Mechanical Systems (CTEMS)* (pp. 114–118).

23. Hiba, A., Hajar, M., Hassan Al, M., & Thomas, N., (2016). Using machine learning algorithms for breast cancer risk prediction and diagnosis. *The 6th International Symposium on Frontiers in Ambient and Mobile Systems (FAMS 2016), Procedia Computer Science* (Vol. 83, pp 1064–1069).

24. Kanchan, P., & Priyanka, C., (2020). Medical Internet of things using machine learning algorithms for lung cancer detection. *Journal of Management Analytics, 7*(4).

25. Radhika, Pr, Rakhi S. Nair, & Veena, G. (2019). A comparative study of lung cancer detection using machine learning algorithms. In: *2019 IEEE International Conference on Electrical, Computer, and Communication Technologies (ICECCT)* (pp. 1–4). ISBN: 978-1-5386-8159-6.

26. Lal, H., Adeel, A., Sharjil, S., Saima, R., Imtiaz, A. A., Saeed, A. S., Abdul, M., et al., (2017). Prostate cancer detection using machine learning techniques by employing a combination of features extracting strategies. *Cancer Biomarkers, 21*(2), 1–21.

27. Sandhya, N. D., (2019). A review on early detection of oral cancer using ML techniques. *International Journal of Scientific Progress and Research (IJSPR), 58*(158), 1. ISSN: 2349-4689.

28. Madhura, V., Meghana, N., Namana, J., Varshini, S. P., & Rakshitha, R., (2019). Survey paper on oral cancer detection using machine learning. *International Research Journal of Engineering and Technology (IRJET)* (Vol. 6, No. 3). e-ISSN: 2395-0056.

29. Prabhakaran, R., & Mohana, J., (2020). Detection of oral cancer using machine learning classification methods. *International Journal of Electrical Engineering and Technology, 11*(3), 384–393.

30. Vijayalakshmi, M. M., (2019). Melanoma skin cancer detection using image processing and machine learning. In: *International Journal of Trend in Scientific Research and Development (IJTSRD)* (Vol. 3, No. 4, pp. 780–784). ISSN: 2456-6470.

31. Sanjana, M., & Hanuman, K. V., (2018). Skin cancer detection using machine learning algorithm. *International Journal of Research in Advent Technology, 6*(12), 3447–3451.

32. Khushboo, M., Hassan, E., Afsheen, A., Fabrizio, F., & Antonello, R., (2019). Cancer diagnosis using deep learning: A bibliographic review. *Cancers, 2019, 11*(1235), 1–36.

33. Wan, Z., Longxiang, X., Jianye, H., & Xiangqian, G., (2020). The application of deep learning in cancer prognosis prediction. *Cancers, 12*(3), 603.

34. Joseph, A. C., & David, S. W., (2006). Applications of machine learning in cancer prediction and prognosis. *Cancer Informatics, 2*, 59–77.

35. Quinlan, J. R., (1986). Induction of decision trees. *Machine Learning, 1*, 81–106.

36. Ripley, R. M., Harris, A. L., & Tarassenko, L., (2004). Non-linear survival analysis using neural networks. *Stat. Med., 23*(5), 825–842.

37. Domingues, I., Pereira, G., Martins, P., et al., (2020). Using deep learning techniques in medical imaging: A systematic review of applications on CT and PET. *Artif. Intell. Rev., 53,* 4093–4160.

38. Kristy, J. B., Catherine, A. F., Haeri, S., Ramya, L. M., Stephanie, D., Eunkyung, A., Dinesh, P., et al., (2012). Advances in the proteomic investigation of the cell secretome. *Expert Rev. Proteomics, 9*(3), 337–345.

39. Gayathri, B. M., Sumathi, C. P., & Santhanam, T., (2013). Breast cancer diagnosis using machine learning algorithms: A survey. *International Journal of Distributed and Parallel Systems (IJDPS), 4*(3).

40. Zilong, H., Jinshan, T., Ziming, W., Kai, Z., Lin, Z., & Qingling, S., (2018). Deep learning for image-based cancer detection and diagnosis—A survey. Pattern Recognition, 83, 134–149.

CHAPTER 15

A Deep Learning-Based Portable Digital X-Ray Devices for COVID-19 Patients

GOPAL SAKARKAR,[1] VIVEK TIWARI,[2] RAMA RAO KARRI,[3] and
SITI SOPHIAYATI YUHANIZ[4]

[1]G. H. Raisoni College of Engineering, Nagpur, Maharashtra, India,
E-mail: gopal.sakarkar@raisoni.net

[2]IIIT Naya Raipur, Chhattisgarh, India, E-mail: vivek@iiitnr.edu.in

[3]Brunei University of Technology, Brunei Darussalam,
E-mail: karri.rao@utb.edu.bn

[4]Universiti Teknologi Malaysia, Johor Bahru, Malaysia,
E-mail: sophia@utm.my

ABSTRACT

The main objective of this research work is to design and improve a portable digital X-ray system to use deep learning approaches to detect COVID-19 symptoms. This integrated and automated system would make it possible to diagnose this disease that affects the human lungs rapidly. This proposed system will be able to detect COVID-19 using X-ray images with deep learning algorithms using natural image classification and visualization capabilities, and integrate the developed AI engine with the portable X-ray device to test and validate the performance of the developed portable X-ray device with the integrated AI engine on actual datasets.

Application of Deep Learning Methods in Healthcare and Medical Science.
Rohit Tanwar, PhD, Prashant Kumar, PhD, Malay Kumar, PhD, & Neha Nandal, PhD (Editors)
© 2023 Apple Academic Press, Inc. Co-published with CRC Press (Taylor & Francis)

15.1 INTRODUCTION

India is the second-largest country after China in population. Out of the total population, the rural population in India was reported at 65.53% in 2019, as per the World Bank collection of development indicators [1]. Evidently, high costly and time-consuming medical treatment and health checkups including blood testing, CT scan, X-rays, MRI are not afford-able for large number of people in India. Also, the use of ambulance/taxi transportation or of personnel pull-out to accompany the patients to and from the hospital is very expensive for poor people of rural areas. But in emergency case, patients has to visit to medical center for either taking treatment or for taking primarily testing. During this visit, any severe patients are much affected with these movements.

The COVID-19 pandemic could be considerably more dangerous in these resource-constrained conditions. Several people have died in 45 African nations, according to the WHO. The true estimates are likely to be significantly higher due to a lack of access to medical care and a scarcity of reverse transcription-polymerase chain reaction (RT-PCR) assays across Africa. In these areas, the policy must place a strong emphasis on detecting and reducing transmission through effective isolation and quarantine procedures [2].

Chest radiography is a rapid and relatively low-cost analytical method that can be used in a number of resource-intensive healthcare settings. Unfortunately, there is a severe lack of radiological expertise in these areas, prohibiting analysis and prediction of such pictures [3]. An AI system can be a useful tool for radiologists or, in the general case where radiological knowledge is not available, for the medical team [4, 5].

Considering all these problems, portable X-ray machine, which will be very handy and easily movable from one place to other as its weight up to 2 kg with integrated AI engine for predict COVID-19 or pneumonia along with the percentage of infection.

COVID-19 is an infectious respiratory disease first identified in late 2019, caused by a novel coronavirus. COVID-19 and its symptoms can vary between mild and severe. Anyone can get COVID-19, but some people are more likely than others to be at risk of serious symptoms. Most people recover within a couple of weeks from COVID-19, but it can be life-threatening also. The best way to diagnose early on and prevent infection is to avoid being exposed to the virus. As the COVID-19 pandemic threatens to overtake healthcare systems worldwide, chest X-rays (CXRs) are suggested as one of the valuable methods to detect COVID-19 at an early stage [6]. A study [7] indicates that of 27% of hospitalized COVID-19 patients, 69% had an irregular chest radiograph at the initial time of admission, and 80% had radiographic anomalies at any time during hospitalization. Around 10–12 days after the onset of symptoms, the results are stated to be the most detailed.

Pneumonia, like COVID-19, is a form of acute respiratory infection that affects the lungs. The lungs are made up of little sacs called alveoli that fill up with air when a healthy person breathes. When a person develops pneumonia, the alveoli get blocked with pus and fluid, making breathing difficult and reducing oxygen intake.

It might be hard to differentiate both pneumonia and COVID-19 since the symptoms are so similar to those of a cold or influenza. CXR is a very effective and efficient method to find out early detection. X-ray imaging is a very costly and lengthy process in India, particularly in rural area. With rising number of COVID-19 patients cases, the number of patients requires. X-ray imaging is also rising making it very difficult and tedious task for doctors.

15.2 THE PROBLEM DEFINITION

The normal X-ray machine used in India, which is very heavy and not portable. Sanitization of machine and room is very cumbersome after each patient because of its huge size and non-portability as shown in Figure 15.1.

Another concern with this system is that an expert technician is required to operate this machine and resultant images are also not available in short time duration. It essentially requires 220 volts; 60 hertz electric supply and a major drawback of this machine is that it does not easily produce digital X-ray images. Reading of X-ray images produced is also not an easy task. Only X-ray images do not highlight the infected areas and hence reading of images requires the presence of an expert.

Despite the fact that a CT scan of the chest provides more reliable features for COVID diagnosis, it is not widely available in resource-constrained settings. Furthermore, CT scan machines are more costly than X-ray machines and can lead to inaccurate diagnosis [8].

FIGURE 15.1   Normal X-ray machine at diagnostic center.

Furthermore, one of the major concerns with this current X-ray machine is that patient has to visit radiologist's center, i.e., X-ray machines are not portable. This is a major hurdle not only for the patient from rural areas and other remote locations but also for the government and to health workers who are unable to cater to such patients.

Considering all the lacunas in the current system and the requirement and comfortability of the patients, it is essential to invent a portable, lightweight AI-based X-ray machine for easy carrying and sending its report to the concerned doctor automatically through e-mails or WhatsApp for further study. India is a developing country, and there is still large scope for improvement in health services, because of current inadequate infrastructure, a smaller number of expert doctors, absent health workers, poor quality, drugs not available, inconvenient hours, long waiting time, or distrust [9]. In darker colored districts, a higher proportion of households cited quality concerns (Figures 15.2 and 15.3).

15.2.1 ORIGINALITY IN RESEARCH WORK

The proposed artificial intelligence (AI) based portable camera size X-ray machine system is very user friendly. As compared to the currently available heavy X-ray machine, this camera is very handy, easy to take imaging. An intelligent AI-based trained model will be installed on existing portable X-ray camera that automatically displays an area of infection with percentage in real-time. Such system is helpful to take primary decisions by unskillful health workers (Figure 15.4).

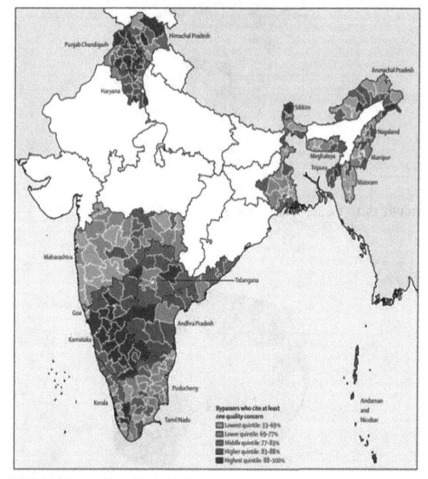

FIGURE 15.2   Health quality and facility sources in India.

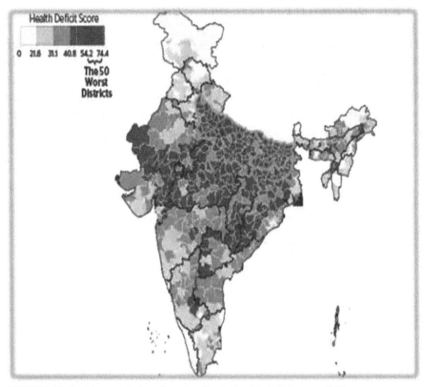

FIGURE 15.3   The map is based on the district health deficit score [10].

FIGURE 15.4   Portable camera size X-ray device.

Another advantage of this device is that it requires DC 5–12 V, 2.1 A, AC 100–240 V, 50–60 Hz with a power output 120 W, its focal spot size 0.4 mm with weight up to 2 kg. To handle this camera, there is no need for an expert technician, and any person can handle it easily, and also, it saves time (Figure 15.5).

FIGURE 15.5   Current system working without AI [11].

Using this handy portable chest X-ray (CXR) system, which takes a chest image of the patient. This camera is connected to the laptop via a Wi-Fi device, and in less than 3 seconds of the image being captured, after the image is stored in the laptop for feature processing in Figure 15.6 (see, https://itie.in/itie-products/portable-digital-x-ray; Table 15.1).

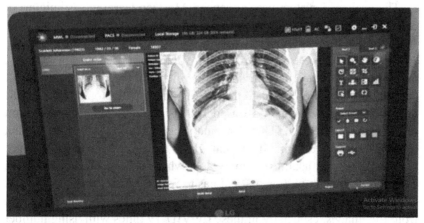

FIGURE 15.6   Chest X-ray image in softcopy format.

TABLE 15.1   Technical Details of Equipment [12]

Name of Product	Digital Portable X-Ray
Focal spot	0.4 mm
Tube kv/mA	Fixed 60 kv/Fixed 2 mA
Expose time	Selectable 0.01~1.30 sec
Frequency	120 khz
Wave dose	0.04 μSv (0.0004 mSv)
Power supply	DC 11.1 V (battery)
Output power	120 W
Expose time	Selectable 0.01~1.30 sec
Input power	X-ray Unit: DC5~12 V, 2.1 A
–	Battery charger: 240 V, 50~60 Hz, 1 A
Battery	1,500 mAh (over 100 exposures per 1 time)
Size/weight	188.5×145×127 mm/1.8 kg

15.3   METHODOLOGY

15.3.1   PROPOSED SYSTEM

In the proposed system, an X-ray image has been taken by a portable camera size X-ray machine as per shown in Figure 15.5. It is then sent to Wi-Fi connected laptop, thereafter if doctor wants to check it, image is readily available in laptop as shown in Figure 15.6.

The proposed system uses advanced AI (deep learning) techniques to predict whether the patient is suffering from COVID-19/Pneumonia. The extent of infection in terms of percentage is also provided by the system.

Figure 15.7 represents the architecture of the proposed system. Initially, portable handy X-ray machine shoots the image of the chest of the patient and it dispatches the images to the Wi-Fi connected laptop automatically. Proposed system exploits AI-Based DL for the automatic classification of the type of infection. This is a binary classification system to predict COVID-19 or pneumonia. It also calculates the percentage of infection.

This system is useful in remote area where doctors are not available 24/7 and, in such condition, patient has to wait for them.

This system also reduces the load on doctors, as it reports the infection from the images, especially in these COVID-19 times with using number of infected portions.

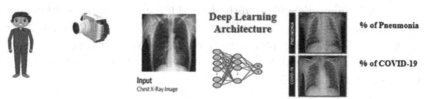

FIGURE 15.7 Proposed system architecture of deep learning-based portable X-ray machine.

15.3.2 PROJECT METHODOLOGY

The research work is planned to follow the methodology as described in Figure 15.8. Next, the data for the research work has to be collected. This is done through data collection through searching from standard benchmark databases, also, data collection through X-ray images taken from the patients in the selected hospitals. The dataset has to be labeled from the experts, and this is where the project collaborators' from India may contribute to the data collection and labeling.

While performing the experiment, the data, which are the labeled X-ray images, will go through a set of preprocessing tasks such as data cleaning, segmentation, transformation, and any other relevant data preprocessing task. The next goal is to construct an AI engine utilizing segmentation. Object detection is a task in computer vision that involves determining the existence, location, and type of one or more objects in a set of X-ray images, with the help of U-Net, SegNet, and CardiacNet architectures that was developed for biomedical image segmentation and ResUnet++ Architecture which is an Advanced Architecture for Medical Image Segmentation. This model could help mitigate the reliability and interpretability challenges often faced when dealing with medical imagery.

The images data included both standard radiographs (posteroanterior and lateral projection) of the chest for testing purposes, of which only the posteroanterior images will be selected. In cases where multiple chest radiographs were available for a patient, the best-quality image acquired for diagnostic purposes will be selected for further AI-based processing and prediction.

Using this system, it will be convenient for doctor to take rapid decision about, patient is suffering from which type of lung infection and because of this portable X-ray machine, it is very useful to move this machine from one place to other. Patient from rural areas traveling to cities for imaging service, can used this system and avoid lengthy hours of traveling and can dispatch to doctor via e-mail or by WhatsApp and after receiving X-ray imaging, doctor has to upload this X-ray image to AI-based engine and it will predict the lung infection.

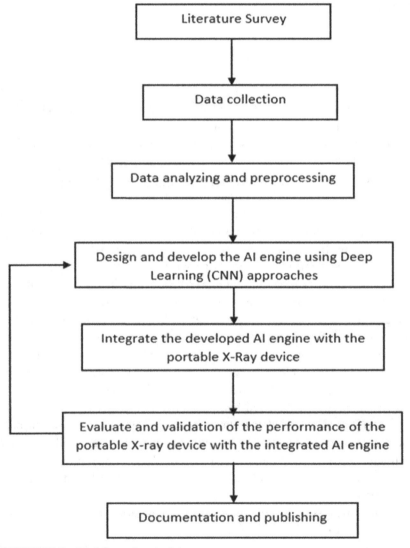

FIGURE 15.8 Workflow of methodology.

The data, which are the labeled X-ray images will then go through a set of preprocessing tasks such as data cleaning, segmentation, transformation, and any other relevant data processing before going to the next task, which is the AI engine development. Segmentation using MRCNN is used for object detection in computer vision which involves identifying the presence, location, and type of one or more objects in given X-ray images, along with U-Net [13], SegNet [14], and CardiacNet [15] architectures that was developed for biomedical image segmentation and ResUnet++ architecture which is an advanced architecture for medical image segmentation. This approach could be able to help with the dependability and interpretability issues that come up frequently when dealing with medical images.

In the AI engine development, the AI engine, or the COVID-19 classifier model will be developed by training the model using the preprocessed datasets in the previous stages. Here well-known pre-trained DL CNNs: AlexNet [16], ResNet, DenseNet, and SqueezeNet [18] will be used to develop the AI engine.

The proposed system involves the introduction of GAIN, CNN, Capsule Network (Cap-sNet), Heap Map for verifying the infection of the lungs area and its level of infection (low, mid, high). Proposed system also employs image enhancement methodology and VGG16 – convolutional network for classifying and detecting infected area of lung.

The proposed system uses the open-source DL architecture of Keras with the backend of TensorFlow to design and train the CNN model. For the purpose of training and research, the standard data set published by the Radiological Society of North America defining the X-ray picture identification and whether pneumonia is present in the X-ray data will be used. The available images are X-rays, which have little color space, so the image does not display a difference on the edges or in the parts where such features can be identified on the RGB scale. Therefore in our picture preprocessing, we have applied some color modifications.

Three separate image processing methods are proposed by our experimental methodology: increase in contrast, extend the color space of the image, and artificially illuminate the image (increase in brightness).

The developed AI engine will then be integrated with the portable X-ray device. There should be issues arising during the integration, and the developed AI engine should be robust enough to be implemented on the portable device. The task on the integration of the AI engine with the portable X-ray device will be jointly working with the Malaysian collaborator. These tasks of AI engine development and integration with the portable device may be iterated, until it reaches the desired performance target.

15.4 RESULT DISCUSSION

Using this system, it will be convenient for doctor to take rapid decision about, patient is suffering from which type of lung infection and because of this portable X-ray machine. Patient from rural areas traveling to cities for imaging service, can used this system and avoid lengthy hours of traveling and can dispatch to doctor via e-mail or by WhatsApp and after receiving X-ray imaging, doctor has to upload this X-ray image to AI-based system, and it will predict the lung infection.

As witnessed by the world, COVID-19 is a dreadful pandemic. It affects the respiratory system and is highly contagious. The only solution to this is distancing. This is where proposed portable digital X-ray devices fill the gap. It will be socially benefitted a lot as it will be highly mobile and lightweight. It avoids transport and extra movements of patients as severe patients are much affected by the movements, which will partially be saved by the proposed system. It will have a high social impact as it will also save the long queues of waiting times required by the traditional system of taking X-rays. The proposed system is convenient for the authorities to detect infection and pneumonia in patients in an easier way. It will reduce hospital overcrowding and save money by reducing the use of ambulance/taxi transportation and personnel pull-out to accompany patients to and from the hospital. All these benefits lead to an increased quality of patient care and sometimes can save more lives. Thus, the proposed system will have greater socio-economic advantages.

15.5 CONCLUSION

After the successful completion of this project, a simple and efficient diagnostic tool will be developed to predict pneumonia or COVID-19 infections. Integration of high-end AI-based techniques with portable X-ray systems will be a boon to the people living in villages and remote areas.

This integrated system will reduce the need of transporting the patients to city hospitals (which normally takes traveling time of 3 to 4 hours for getting an X-ray image), early detection of the infection, get a quick and efficient diagnosis of patient health from a well-established doctor from a government or premium hospital (digital X-ray along with machine-based pre-diagnostic results can be sent to the doctor via e-mail). This proposed system not only saves valuable time, but also can save human lives due to early and accurate detection of infection.

Therefore, this system will have a huge social impact on society and minimize the risk of spreading contagious diseases.

KEYWORDS

- artificial intelligent
- COVID-19
- deep learning
- reverse transcription-polymerase chain reaction
- X-ray images

REFERENCES

1. https://tradingeconomics.com/india/rural-population-percent-of-total-population-wb-data.html (accessed on 14 May 2022).
2. Keelin, M., Henk, S., Arnoud, J. G. K., Michael, B. J. M. K., Sam-Son, T., Ernst, T. S., Steven, S., et al., (2020). COVID-19 on chest radiographs: A multireader evaluation of an artificial intelligence system. *Radiology, 296*(3).
3. Mollura, D. J., Culp, M., & Lungren, M. P., (2019). *Radiology in Global Health, Strategies, Implementation, and Applications* (2nd edn.). Cham, Switzerland: Springer.
4. Hwang, E. J., Nam, J. G., Lim, W. H., et al., (2019). Deep learning for chest radiograph diagnosis in the emergency department. *Radiology, 293*(3), 573–580.
5. Annarumma, M., Withey, S. J., Bakewell, R. J., Pesce, E., Goh, V., & Montana, G., (2019). Automated triaging of adult chest radiographs with deep artificial neural networks. *Radiology, 291*(1), 196–202.
6. Arpan, M., Surya, K., Harish, R., Krithika, R., Vinay, N., Subhashis, B., & Chetan, A. (2020). *CovidAID: COVID-19 Detection Using Chest X-ray.* https://arxiv.org/abs/2004.09803.

7. Wong, H. Y., Lam, H. Y., Fong, A. H., Leung, S. T., Chin, T. W., et al., (2020). Frequency and distribution of chest radiographic findings in COVID-19 positive patients. *Radiology*, 201160.
8. Sarkodie, B. D., Osei-Poku, K., & Edmund, B., (2020). Diagnosing COVID-19 from chest X-ray in resource-limited environment-case report. *Medical Case Reports-2020* (Vol. 6, No. 1, p. 135). ISSN: 2471-8041.
9. https://www.indiaspend.com/more-indians-die-of-poor-quality-care-than-due-to-lack-of-access-to-healthcare-1-6-million-64432/ (accessed on 14 May 2022).
10. https://www.livemint.com/Politics/aubiq1yR8CiYZ6G9T7NWiP/Ayushman-Bharat-health-insurance-dstricts.html (accessed on 14 May 2022).
11. https://youtu.be/WJpF9uyAY4s (accessed on 14 May 2022).
12. https://itie.in/itie-products/portable-digital-x-ray/ (accessed on 14 May 2022).
13. Olaf, R., Philipp, F., & Omas, B., (2015). *U-Net: Convolutional Networks for Biomedical Image Segmentation*. MICCAI Springer, New York, NY, USA.
14. Badrinarayanan, V., Kendall, A., & Cipolla, R. (2017). "SegNet: A Deep Convolutional Encoder-Decoder Architecture for Image Segmentation," In *IEEE Transactions on Pattern Analysis and Machine Intelligence*, 39(12), 2481–2495, 1 Dec. 2017, doi: 10.1109/TPAMI.2016.2644615.
15. Aliasghar, M., Rashed, K., Kawal, R., Jeremy, B., & Ulas, B., (2017). *CardiacNET, Segmentation of Left Atrium and Proximal Pulmonary Veins from MRI Using Multi-View CNN*. MICCAI, Springer, New York, NY, USA.
16. Krizhevsky, A., Sutskever, I., & Hinton, G. E., (2012). ImageNet classification with deep convolutional neural networks. In: *Advances in Neural Information Processing Systems* (pp. 1097–1105).
17. LeCun, Y., Kavukcuoglu, K., & Farabet, C., (2010). Convolutional networks and applications in vision. In: Proceedings of 2010 IEEE International Symposium on Circuits and Systems.

CHAPTER 16

Adoption of Machine Learning and Open Source: Healthcare 4.0 Use Cases

NEELU JYOTHI AHUJA

Department of Systemics, School of Computer Sciences, University of Petroleum and Energy Studies, Dehradun, Uttarakhand, India, E-mail neelu@ddn.upes.ac.in

ABSTRACT

Application of artificial intelligence (AI) and precisely, the machine learning algorithms have seen rapid acceleration in the healthcare domain, in recent times. It has fueled transformative trend, focused on improved patient outcomes. Machine learning algorithms learn from data independently without explicit human intervention, and assist decision-making. In other words, they have the potential to ingest large medical datasets to discover data patterns and deliver complex decision-making. Studying thousands of healthcare records and other patient data, these can detect patterns associated with diseases and health conditions. However, challenges of the requirement of large volumes of training data sets, adding on to expenses need attention. Open source tools provide economic solution. The present work discusses the application of class of machine learning and deep learning (DL) algorithms, with open-source data science tools, in offering healthcare solutions with higher accuracy and precision, across the spectrum of successful diagnosis to recommendation of precise treatment. Five use cases of healthcare 4.0 presented with recommendations, advocate, active application of machine learning, towards patient-centric approach, targeted to enhanced patient outcomes and experience.

Application of Deep Learning Methods in Healthcare and Medical Science.
Rohit Tanwar, PhD, Prashant Kumar, PhD, Malay Kumar, PhD, & Neha Nandal, PhD (Editors)
© 2023 Apple Academic Press, Inc. Co-published with CRC Press (Taylor & Francis)

16.1 INTRODUCTION

Deep learning (DL), a complex type of machine learning is widely used in several aspects of medical science and healthcare. Some of them are radiology, medical imaging (MI), wherein, the DL algorithms learn to identify cancerous lesions from images. In addition, detection of any abnormalities, musculoskeletal injuries, etc., is possible. Identification facilitates an early treatment.

Augmented reality (AR), virtual reality (VR), collectively referred extended reality (ER), 3D printing, and wearable technology, are amongst the top technologies proposed to have tremendous impact on healthcare in the near future. As more and more people embracing wearable technology in form of smartwatches and fitness trackers, generates large volumes of data, which when ingested in DL algorithms reveals valuable insights. Application of nanotechnology, in medicine, referred to as nanomedicine, is execution of tasks such as drug delivery at nanoscale, i.e., at molecular/cellular/DNA level. In this case, drug delivery is at the targeted region, enabling site-specific action, bypassing healthy cells/tissues.

Latest developments include the use of machine learning algorithms for diagnosis of diabetic retinopathy (DR) and predictive analytics to determine reoccurrence of breast cancer, basis analysis of image data sets. DL based prediction models developed at Massachusetts Institute of Technology help to predict breast cancer in its early stages.

Electronic health record (EHR) is an electronic version of the medical history of a patient, which includes all essential details of the patient, including demography, treatment details, vital information related to health as maintained by the provider. AI supports EHR management, through improved recordkeeping, facilitating enhanced patient care. Clinical decision making greatly depends on the data fed to the machine learning algorithms, hence, data preprocessing, that usually involves data gathering, analyzing, classifying, and cleansing of data, has an important role in maintaining data integrity. Machine learning combined with predictive analytics paves the way for improved patient outcomes, leading to near accurate diagnosis and treatment, assisting treating physicians with helpful insights for personalized treatment.

Machine learning facilitates precision and accuracy in the use of surgical robotic tools, based on real-time analysis of data of previous surgeries, and past medical records and assists complex surgeries such as unclogging blood vessels of heart, and spine surgery through well-planned workflows.

Personalized medicine, also referred to as precision medicine, is a model that categorizes patients into different groups, based on their risk to a disease, or predicted response to a particular type of treatment. Accordingly, to offer treatment, intervention, administering products and services, specifically tailored to them. Machine learning algorithms apply analytics on large volumes of patient data and facilitate the development of personalized treatments. ML helps to harness data from EHRs and acts as a valuable decision-making tool, to support, in cases where patients are incapable of making independent decisions. Reduced time to review patients leading to faster diagnosis, earliest treatment and faster recovery are advantages of DL algorithms. ML has been useful in predicting COVID-19 surges in times of global pandemic.

Machine learning enables analysis of genetic mutations fast. ML applications facilitate genome sequencing that can help to predict risk for a particular disease, to an individual. ML has made genome sequencing more affordable.

One of the challenges of machine learning is it requires large volumes of data, for training and validating algorithms. The larger the data set, the better is precision in the task of the algorithm. For example, trained on a large volume of data set, a predictive model is likely to deliver precise prediction, in a task such as identification of malignant tissues and recommending appropriate patient care. Large dataset leads to higher cost. Open source offers several advantages, such as high-quality software made available without vendor lock-in, allowing for experimentation and custom building of applications, to meet stringent regulatory requirements and patient care standards. Open source provides healthcare businesses, to experiment and adapt models addressing a variety of challenges, including the cost of training data. This gives a competitive advantage to the firm, paving a way to accelerated pace of scientific discovery, improved human insight and enhancing medical practices.

16.2 RELATED WORK

As reported in Imaging Technology News, the estimate is that the growth in AI in the healthcare sector will rise by 40% by 2025, as against its share in 2018.

Open source: Developing and evaluating machine-learning algorithms using open source tools is very widely used in current times. Sidey-Gibbons and Sidey-Gibbons [1] demonstrated the use of machine-learning techniques, General Linear Model Regression (GLMs), Support Vector Machines (SVMs) with radial basis function kernel, and single layer ANNs, in development of predictive models for detection of breast cancer. The authors used public domain dataset, breast cancer Wisconsin Diagnostic data set, describing breast mass samples, and an open-source R statistical environment for coding. R environment is accessible through R Studio, which is open source integrated developer environment.

Another challenge is that most of the healthcare solutions solve a particular problem, such as identifying a lesion or malignant tissue or risk of developing of a sepsis. This makes it difficult to customize the underlying models to deliver the best as against the investment. Thus, the present work recommends the use of open-source data science tools to adapt models for different problems in healthcare and medical science.

With the coming together of large electronic data sources and computing power, the efficiency of AI and machine learning algorithms lead to successful identification of clinically meaningful patterns.

Shah et al. [2] presented a translational approach for clinical development and healthcare, recommending confluence of technology corporations, regulators, non-profit organizations, academics, and industry, particularly biotechnology industry. The authors emphasized, actionable computational evidence, for clinical development and healthcare, as an outcome of this approach. They highlighted the impact of algorithmic evidences in patient care, and outlined the improvement in clinical development process through integration of AI and ML-based technologies.

While machine-learning tools deliver meaningful insights and assist clinical decision making, translating the needs of healthcare, to leverage theoretical achievements in realistic settings is a challenge, yet to successfully conquer. Mateen et al. [3] emphasized on the need for synergistic efforts of experts from both machine learning research and healthcare. The authors recommended the extension of reporting guidelines in clinical and health sciences to include machine-learning approaches. They presented examples of checklists for reporting clinical trials involving AI and related methodologies, seeking review and discussion from researcher communities.

Kelly et al. [4] outlined applications of AI integration in healthcare research and referring to it as transformative technology, explored the potential of expansion across domain of medicine. The reported real-time challenges in deployment of these techniques in clinical practice. They highlighted the limitations and recommended road map to translate the technologies from research to clinical practice.

Sidey-Gibbons and Sidey-Gibbons [5] highlighted the need for capacity development in the area of predictive machine learning algorithms supporting decision-making in medical and clinical research. They advocated the use of open-source software and publicly available data sets to pursue the development and evaluation of predictive algorithms, along with enumeration of available resources and guided steps.

Shailaja, Seetharamulu, and Jabbar [6] discussed the significance of machine learning in disease diagnosis. They presented a review of different machine learning algorithms used in decision-making in healthcare. The authors proposed methodology for design and development of efficient clinical decision support system for medical domain.

Senerath and Gamage [7] presented a detailed review of the use of machine learning and DL algorithms in decision making across different aspects of healthcare, MI, precision medicine and computational biology.

Vollmer et al. [8] explored the potential of the use of AI and statistical methods in discovering valuable insights from voluminous patient data. They presented concerns around lack of transparency, clear reporting facilitating replicability, ethics, and effectiveness, particularly in the case of imaging data. They presented 20 critical questions around these concerns and recommended a guide of best practices to leverage the benefits of machine learning.

Section 16.3 presents five use cases for machine learning in the healthcare sector, focused towards improvement of patient outcomes.

16.3 HEALTH 4.0 APPROACH FOCUSED ON IMPROVING PATIENT OUTCOMES: USE CASES

As defined by WHO, an outcome is a change in the health of an individual or a group of individuals or population, attributable to an intervention or a series of interventions. The measures being mortality, readmission, and patient experience.

1. **Support to Administrative Tasks Through Natural Language Processing:** Physician burnt out is one of the problems, where the physician is occupied with administrative tasks to the extent as high as 83% (as reported by the New England Journal of Medicine) of the time, leading to inability to spend time with patients. This affects the patient experience. The recommendation is offloading of administrative tasks. A good amount of administrative tasks involve the review and update of EHRs. Natural Language Processing algorithms identify and categorize words and phrases. This facilitates a physician to dictate notes directly, to EHRs during the patient interaction, allowing for more time spent in actual patient consultation, than in administrative recording. This leads to improved patient experience. NLP tools allow for neatly compiled summaries and charts with well-augmented evidence, to review the health of the patient. By leveraging NLP, the physician can save time spent on comprehending test results, and maintaining EHRs, spending it productively with the patient, leading to enhanced outcomes.

2. **Anomaly Detection in Patient Data for Risk Prediction:** Applying machine learning models with anomaly detection algorithms on EHRs assists medical practitioners and healthcare professionals to uncover the risk of heart attacks, strokes, sepsis, or any other serious ailments. EHR serves as a source of data containing patients' historical data, daily evaluation summary, real-time measures of vital parameters, heart rate variability data, and blood pressure-related data. Timely discovery and alert assist in taking prompt preventive actions. The case in focus is the development of a tool by El Camino Hospital to predict patient falls by regular monitoring of EHR, bed alarm, and nurse call data. The tool categorizes patients into high, medium, and low-risk categories. Recent data from the tool reveals the alert serves to reduce falls by 39%. Apart from reduced trauma, the reduction in patient fall is directly attributable to saving expense and hospital stay period.

   A Sepsis Sniffer Algorithm developed by Mayo Clinic uses data of demography, and vital sign measures, to trigger alerts when identifying the risk of developing sepsis. The use of machine learning reveals results as faster as 72% times the traditional method of identification. This once again is the direct implication, allowing for maximizing of time of the doctor, with the patient, thus, generating improved patient outcomes.

3. **Leveraging Data Visualization for Insights to Accelerate Medical Research:** With an overwhelming number of research articles, studies, and reports, and social media comments, available across medical research and other interdisciplinary fields, skimming out valuable insights is a challenge for scientists and practitioners. The use of DL algorithm-based predictive tools, text mining, and NLP tools to parse through and analyze literature is one of the solutions, further presented through data visualization tools is one of the potential solutions. A study undertaken by the US and Ireland jointly on Adverse Drug Events analyzed over 3,000,000 articles using text mining, and predictive analytics and presented relationships between drugs and side effects, through data visualization tools. EHR data is largely unstructured, as it involves a variety of non-computer-based formats such as physicians' notes, ECG results, etc. NLP is useful for mining unstructured data and extract useful insights.

4. **Assistance to Radiologist for Precise Interpretation of Radiographic Images:** A radiologist, on average, interprets a radiographic image every three seconds. Monotonous work and an increased degree of ambiguity may affect their interpretation at times. In addition, continuous fatigue sets in leading to imprecise interpretation. DL algorithms trained on radiographic image data set, assist radiologists in ascertaining and recognizing the development of tumors at early stages. Thus, the recognition of tumors by analysis of complex patterns in radiographic image data to recognize tumors of lungs, breasts, brain, or any other area with good accuracy, paves the way for improved patient care and outcomes. ML-based diagnostic tools developed by Houston Methodist Research Institute for breast cancer detection interpret mammograms with more than 99% accuracy.

5. **CNN-Based Skin Cancer Diagnosis:** Using machine learning resources such as TensorFlow, sci-kit-learn, and Keras, CNN-based tools developed assist recognition and classification of images, and detection of skin cancer, up to an accuracy of closer to 90%. This is higher than the accuracy rate of dermatologists detecting

melanomas. Dataset with a voluminous number of images of malignant and benign skin lesions available as a public domain resource is valuable for training and fine-tuning the machine-learning models. Several projects developed by academia, predominantly using open-source tools, uploaded on repositories such as GitHub serve as a valuable resource for future research and development. The use of CNN-based tools for the diagnosis of a number of diseases such as Alzheimer's disease (AD), CVDs, and tuberculosis is widely available in recent times.

Privacy, security, governance, and compliance are the key requirements of the healthcare industry, as it deals with real time-sensitive medical data. The training and testing of models on anonymous data sets, complying with Health Insurance Portability and Accountability Act of 1996 (HIPAA) requirements, practiced renders models that deliver high accuracy, ultimately aimed at improved patient outcomes.

16.4 CONCLUSION

Application of a wide range of classes of machine learning models in clinical, medical, and healthcare sectors is at risk, delivering benefits to society in terms of reducing the disease burden by assisting diagnosis and early treatment. Five use cases of healthcare 4.0 discussed here contribute to building a knowledge base of current developments in machine learning models, particularly predictive models, introducing their utility in the direction of patient care, improved patient outcomes, and enhanced experience. However, the aspect of adherence to the regulatory framework and its role in use is not in the context of the present work.

KEYWORDS

- **augmented reality**
- **electronic health record**
- **general linear model regression**
- **support vector machines**
- **virtual reality**

REFERENCES

1. Sidey-Gibbons, J., & Sidey-Gibbons, C., (2019). Machine learning in medicine: A practical introduction. *BMC Med. Res. Methodol., 19*, 64. https://doi.org/10.1186/s12874-019-0681-4.
2. Shah, P., Kendall, F., Khozin, S., et al., (2019). Artificial intelligence and machine learning in clinical development: A translational perspective. *Npj Digit. Med., 2*, 69. https://doi.org/10.1038/s41746-019-0148-3.
3. Mateen, B. A., Liley, J., Denniston, A. K., et al., (2020). Improving the quality of machine learning in health applications and clinical research. *Nat. Mach. Intell., 2*, 554–556. https://doi.org/10.1038/s42256-020-00239-1.
4. Kelly, C. J., Karthikesalingam, A., Suleyman, M., et al., (2019). Key challenges for delivering clinical impact with artificial intelligence. *BMC Med., 17*, 195. https://doi.org/10.1186/s12916-019-1426-2.
5. Sidey-Gibbons, J., & Sidey-Gibbons, C., (2019). Machine learning in medicine: A practical introduction. *BMC Med. Res. Methodol., 19*, 64. https://doi.org/10.1186/s12874-019-0681-4.
6. Shailaja, K., Seetharamulu, B., & Jabbar, M. A., (2018). Machine learning in healthcare: A review. In: *2018 Second International Conference on Electronics, Communication, and Aerospace Technology (ICECA)* (pp. 910–914). doi: 10.1109/ICECA.2018.8474918.
7. Senerath, M. D. A. C. J., & Gamage, U. G., (2021). Involvement of machine learning tools in healthcare decision-making. *Journal of Healthcare Engineering, 2021*, 20. Article ID: 6679512, https://doi.org/10.1155/2021/6679512.
8. Vollmer, S., Mateen, B. A., Bohner, G., KirÃ¡ly, F. J., Ghani, R., Jonsson, P., et al., (2020). Machine learning and artificial intelligence research for patient benefit: 20 critical questions on transparency, replicability, ethics, and effectiveness *BMJ, 368*, l6927. doi: 10.1136/BMJ.l692.

Index

X-ray
  diagnostic models, 190
  images, 124–126, 129, 132, 135, 142,
    244, 259, 265, 267, 273, 275, 277
  imaging, 155, 267, 274, 276
    strategies, 155
  photograph, 247
  print photograph, 247

Z

Zero temperature, 153
Zooming, 131, 134
Zoonotic diseases, 185

Printed in the United States
by Baker & Taylor Publisher Services